William White Cooper

Curious and Instructive Stories about Wild Animals and Birds

William White Cooper

Curious and Instructive Stories about Wild Animals and Birds

ISBN/EAN: 9783337231286

Printed in Europe, USA, Canada, Australia, Japan

Cover: Foto ©berggeist007 / pixelio.de

More available books at **www.hansebooks.com**

CURIOUS AND INSTRUCTIVE STORIES

ABOUT

WILD ANIMALS AND BIRDS

"The study of the works of Nature is the most effectual way to open and excite in us the affections of reverence and gratitude towards that Being whose wisdom and goodness are discernible in the structure of the meanest reptile."—IZAAK WALTON.

EDINBURGH
W. P. NIMMO, HAY, & MITCHELL
1897

CONTENTS.

CHAPTER I.

Invisibility of Lions.—Their Eyes.—Stormy Nights.—A cool Lion.—Gregarious Spirit.—Mr. Gordon Cumming.—Voice of Lion.—A pleasant Party.—Three jolly Lions.—First Menagerie.—The Lion Tower.—"The Lions."—Anecdote.—A Sick Lion.—Sir E. Landseer and his early Visitor.—Ancient Remedy for an Invalid.—The Tongue in the Felinæ.—Stillness.—The Lion's Tail.—Man-eaters.—Fearful Death.—Adventure of Captain Woodhouse.—The Gentle Lion.—A Noble Lion.—The Retreat of the Leonidæ.—How to Face a Lion.—An Adventure.—Forbearance of the Lion.—Mungo Park.—An Affectionate Captive.—An Adventure.—Strength of Lions.—Pumas.—Ferocity of Puma.—Edmund Kean's Puma.—Zoological Gardens . . 1

CHAPTER II.

Bears in Britain.—Our Ancestors and their Bears.—Bear-baiting. — Zoological Gardens. — The Grizzly Bear. — Strength of Grizzly Bear.—Anecdote.—Adventure.—Ursine Interments.—Shooting Grizzly Bears.—Bears at the Gardens. — Operations for Cataract. — A Chloroformed Bear.—Operation on another Bear.—Arctic Expedition.—A Crafty Intruder.—Lapland Hunters.—Mutual Politeness.—The two Friends.—An Ill-assorted Couple.—An Adventure.—A Narrow Escape.—Swedish Skalls.—An Accident.—Jan Svenson.—The Bear and the Chasseur.—A Scalp.—An Awful Situation.—The General and the Bear.—Wolves and Bears.—A Swiss Myth.—The Oxford Bear.—Tiglath-pileser.—An Ursine Undergraduate.—A Bear in a Bed-room.—An Unexpected Visitor.—Tig on Horseback.—A Bear among the Savants.—A Broken Heart 35

CHAPTER III.

Adaptation of Colours.—Highland Tartans.—Varieties.—
Tempers.—Tame Leopards.—The Milliners' Friends.—
Fair Retaliation.—Anecdote of a Panther.—Saï and his
Keeper.—An Alarm.—Affection of a Panther.—Saï and
the Orang.—Short Allowance.—Mr. Orpen's Encounter.—
Tenacity of Life.—A Conflict.—Major Denham.—Lurking
Panthers.—Trapping a Leopard.—South African Leopard.
—Pleasant Surprise.—Tree Tigers.—Eager Sportsmen.—
The Tree Tiger and Artillerymen.—A Practical Joke.—
Scenery of the Orinoco.—An Agreeable Neighbourhood.—
Jaguar and Vultures.—Al Fresco Troubles.—A Bold
Thief.—Jaguar and Turtle.—Mode of Attack.—Sacrilege.
—Favourite Trees.—Leopards in Trees.—The Guacho
and the Leopard.—Narrow Escape.—Poisoned Arrows.—
A Rough Playfellow.—Advantage of Politeness.—Value
of Skins 78

CHAPTER IV.

Wolves.—Origin of Dogs and Wolves.—Points of Difference.—Varieties. — Ancient Superstitions. — Myths. —
Charms.—Wolves in Britain.—Ancient Laws.—The Last
Wolf. — Cunning. — An Unwelcome Visitor. — A Clever
Performer.—A Sociable Animal.—Value of Skins.—Cross
Breed.—An Adventure.—Traps.—Wolf and Reindeer.—
Boldness.—Dogs killed by Wolves.—Cunning of Foxes.—
A Midnight Struggle.—An Affectionate Wolf.—Tussa.—
Theft of an Infant.—Taste for Pork.—An Awkward Predicament.—A Fierce Pig.—A Soldier devoured.—Cry of
the Jackal.—The Old Quarter-master.—An Attractive Figure.—The Sailor and the Beef.—Utility of Wolves . . 126

CHAPTER V.

Antiquity of Horsemanship.—Bucephalus.—British War-chariots.—Aboriginal Ponies.—Stone Horse-collars.—First

CONTENTS. xi

Page

Racers.—Master of the Horse.—Arabians.—Lord Burghley.—Lord Herbert of Cherbury.—Speed of Horses.—Godolphin Arabian.—Eagerness of Horses.—An Old Warrior. —A Surprise.—First Horse in America.—A Domidor.— Breaking a Horse.—South American Steeds.—Turning the Tables.—Swimming.—Turkuman Horses.—A Heroine.— Abd-el-Kadir.—Arab Maxims.—The Warriors of the Desert.—A Fight.—The Defeat.—A Complimentary Excuse. —A Nice Distinction.—Mungo Park and his Horse.—A Sad Loss.—A Noble Animal.—Robbers of the Desert.—A Pet.—The Camanchees.—"Smoking" Horses.—"Charley."—The Faithful Steed.—Wild Horses.—A Bold Stroke. —A Gallant Charger.—Corunna.—Nebuchadnezzar . . 153

CHAPTER VI.

New Species.—Ancient Knowledge of the Giraffe.—Pliny. —Lorenzo the Magnificent.—Giraffe at Paris.—George the Fourth.—Capture of Giraffes—Singular Procession.—Great Attraction.—The First Birth.—Accident.—A Dainty Dish. —Tongue of Giraffe.—Petty Larceny.—The Peacock untailed.—Movements of Giraffe.—Gentleness.—Sir Cornwallis Harris.—First View.—A Disappointment.—Vexatious Incidents.—Coolness of Hottentots.—Another Troop. —Hostile Rhinoceros.—Hot Chase.—The Death.—A Deserter.—Adaptation of Form.—Reflections 197

CHAPTER VII.

Rapacious Birds.—The Eagles of Antiquity.—Cæsar's Standard-bearer.—Waterloo.—The Death of Æschylus.—The Maid of Sestos.—The Phœnix.—Flight of the Eagle.— Gluttons.—Miserable Death.—Anecdote.—Golden Eagle. —The Eagle of Westminster.—A Nocturnal Alarm.—A Tyrant.—An Uninvited Guest.—An Escape.—Tame Eagle.— A Useful Neighbour.—A Mountain Larder.—The Cat-Killer.—Hunting in Couples.—The Invocation.—American

Indians.—Eagles and Reindeer.—The Garter.—Skua Gull.
—Crest of the Stanleys.— The Eagle's Nest.—The Child
Stealer.—A Brave Lad.—A Battle with a Turbot.—Acute
Vision.—Bald Eagle.—Wilson the Poet.—Niagara.—Benjamin Franklin.—The Valiant Thomas.—The Devoted
Parents.—Australian Eagles.—The Future 229

CHAPTER VIII.

Spaniards in Mexico.—Ancient Mexicans.—Feather Embroidery. — Montezuma's Aviary. — Gorgeous Array.—
Works of Art.—Distribution of Humming Birds.—Rapidity of Flight.—Nests.—Courting.—Singular Bower.—
Actions of Polytmus.—Nest-making.—Anecdote.—Mode
of Capture.—Witty Epitaph.—The Philosopher and the
Middies.—Pets.—Trochilus in Captivity.—Song.—Bluefields Ridge.—Gorgeous Scene.—Pugnacity of Humming
Birds. — A Combat. — The "Doctor Bird." — Favourite
Resort.—Tameness—A Bold Bird.—Interesting Birds.—
Notes for Ornithologists.—Tongue of Humming Bird.—
Birds and Spiders.—Mode of Taking Food.—Mr. Gould's
Collection.—Elegant Arrangement.—'The Trochilidæ.'—
Domestication 268

CHAPTER IX.

Crocodiles.—Ancient Writers.—Mode of Capture.—Sacred
Crocodiles.—Tentyrites.—Rare Book.—Indian Worship.
—Medicinal Virtues.—Crocodiles and Alligators.—Anatomical Peculiarities.—Teeth.—Nidification.—Crocodile and
Trochilus.—The Ziczac.—Crocodile Bird.—Hybernation.—
Jacarés of the Amazon.—Poachers.—Mr. Spruce.—Anecdote.—Search for Victoria Regia.—A Disagreeable Neighbour.—The Battle.—The Death.—A Dainty Luncheon.—
Alligators and Dogs.—Mr. Waterton.—Riding on Crocodiles.- A Bold African.—Alligator Tank.—The Subaltern's
Sport.—Conclusion 304

CURIOUS AND INSTRUCTIVE STORIES

ABOUT

WILD ANIMALS AND BIRDS.

CURIOUS AND INSTRUCTIVE STORIES

ABOUT

WILD ANIMALS AND BIRDS.

CHAPTER I.

INVISIBILITY OF LIONS.—THEIR EYES.—STORMY NIGHTS.—A COOL LION.—GREGARIOUS SPIRIT.—MR. GORDON CUMMING.—VOICE OF LION.—A PLEASANT PARTY.—THREE JOLLY LIONS.—FIRST MENAGERIE.—LION TOWER.—"THE LIONS."—ANECDOTE.—A SICK LION. —SIR E. LANDSEER AND HIS EARLY VISITOR.—ANCIENT REMEDY FOR AN INVALID.—THE TONGUE IN THE FELINÆ.—STILLNESS.— THE LION'S TAIL.—MAN-EATERS.—FEARFUL DEATH.—ADVENTURE OF CAPTAIN WOODHOUSE.—THE GENTLE LION.—A NOBLE LION.— THE RETREAT OF THE LEONIDÆ.—HOW TO FACE A LION.—AN ADVENTURE.—FORBEARANCE OF THE LION.—MUNGO PARK.—AN AFFECTIONATE CAPTIVE.—AN ADVENTURE.—STRENGTH OF LIONS. —PUMAS.—FEROCITY OF PUMA.—EDMUND KEAN'S PUMA.—ZOOLOGICAL GARDENS.

"Thou makest darkness, and it is night, wherein all the beasts of the forest do creep forth. The young lions roar after their prey, and seek their meat from God.

"The sun ariseth, they gather themselves together, and lay them down in their dens."—PSALM civ.

IT would be difficult to find language which more simply and elegantly describes the habits of the Mon arch of the Forest than these words of the Psalmist, and they are strictly in accordance with truth. During the day, the lion lies concealed beneath the shade of some thick, stunted tree, or buries himself in a covert of lofty reeds or thick grass; but when the sun goes down, and the shades of evening fall, he sallies forth to prowl during the hours of night. The tawny colour of his hide is admirably adapted for his concealment. Mr. Gordon Cumming (whose remarkable work* contains the best information on the habits of the South African wild animals,) states that he has often heard lions lapping water at a less distance from him than twenty yards, and, although blessed with the keenest vision, he was unable to make out even the outlines of their forms. Their eyes, however, glow like balls of fire, which may be thus explained. In many animals, the inner surface of the back of the eye presents a membrane called tapetum lucidum, which, in the lion and cat tribe, is of a yellow colour and brilliant metallic lustre, like a concave mirror; it is the reflection from this which causes the " glare" of their eyes; thus are they peculiarly fitted for nocturnal habits, but ill-adapted to bear strong sunlight. Some travellers have described what would certainly appear, at first sight, to have been cowardly retreats on the part of lions; but

* A Hunter's Life in South Africa.

doubtless, in the majority of instances where they have turned tail on inferior antagonists, they were conscious of the disadvantage under which they laboured from their eyes being dazzled by the intense glare of an African sun reflected from the burning sands of the desert.

It is on dark and stormy nights that

"Through the gloom,
Loading the winds, is heard the hungry howl
Of famished monsters."

Then it is that the lions, like the witches of old, hold their hideous revels! Then does it behove the traveller to watch with unceasing vigilance, and, if in a district populous with lions, he may esteem himself fortunate should he escape with minor losses. The sentry, as he walks his round, runs much risk of being carried off:

"And while his thoughts oft homeward veer,
A well-known voice salutes his ear,"

in the terrific and heart-paralyzing roar with which the lion springs upon his prey. The journal of the Landrost Jah. Sterneberg, affords a painful example of such a calamity.

"The waggons and cattle had been comfortably put up for the night, when about midnight they got into complete confusion. About thirty paces from the tent stood a lion, which, on seeing us, walked very deliberately about thirty paces further behind a small thorn-bush, carrying something with him, which I took to be a young ox. We fired more than sixty shots at the bush. The south-east wind blew strong, the sky was clear, and the moon shone very bright, so that we could perceive anything at a short distance. After the cattle had been quieted again, and I had looked over everything, I missed the sentry from before the tent. We called as loudly as possible, but in vain; nobody answered, from which I concluded he was carried off. Three or four men then advanced very cautiously to the bush, which stood right opposite to the door of the tent, to see if they could discover anything of the man, but returned helter-skelter, for the lion, who was still there, rose up, and began to roar. About a hundred shots were again fired at the bush, without perceiving anything of the lion. This induced one of the men again to approach it with a firebrand in his hand; but as soon as he approached the bush, the lion roared terribly, and leaped at him, on which he threw the firebrand at him, and the other people having fired about ten shots at him, he returned immediately to his former station; the firebrand, which he had thrown at the lion, had fallen into the midst of the bush, and, favoured by the wind, it began to burn with a great flame, so that we could see very clearly into it and through it. We continued our firing into it; the night passed away and the day began to break, which animated every one to fire at the lion, because he could not lie there without exposing himself entirely. Seven men, posted at the furthest waggons, watched to take aim at him as he came out; at last, before it became quite light, he walked up the hill with the man in his mouth, when about forty shots were fired without hitting him." The end was, that he made his escape in perfect safety. In this narrative it is hard to say which is most to be marvelled at, the wonderfully bad shooting of the men, or the cool, dogged obstinacy of the lion. He seemed to be quite aware of the sort of men he had to deal with, and to have diverted himself with their fears. Not less than three hundred shots must have been fired at him, and yet unscathed he carried off the wretched man.

The late Sydney Smith, in his witty and able 'Sketches of Moral Philosophy,' thus argues, when comparing mankind with the brute creation:—"His gregarious nature is another cause of man's superiority over all other animals. A lion lies under a hole in a rock, and if any other lion happen to pass by, they fight. Now, whoever gets a habit of lying under a hole in a rock, and fighting with every gentleman who passes near him, cannot possibly make any progress. . . . If lions would consort together, and growl out the observations they have made about killing sheep and shepherds, the most likely place for catching a calf grazing, and so forth, they could not fail to improve." Unfortunately for the argument, it was based upon a fallacy, for the observations of Mr. Cumming prove that lions live and hunt in troops, and, for aught we know, may benefit by that very gregarious spirit which the worthy Canon imagined them to want. "It is a common thing," says Mr. Cumming, "to come upon a full-grown lion and lioness associating with three or four large young ones, nearly full-grown. At other times, full-grown males will be found associating and hunting together in a happy state of friendship; two, three, and four, may thus be discovered consorting together." To that intrepid sportsman, the grandest music was the roar of troops of lions, as three or four of these advanced from different quarters to the same watering-place, and no description could more accurately convey an idea of this terrible though sublime sound than this.

"One of the most striking things connected with the lion is his voice, which is extremely grand, and peculiarly striking. It consists, at times, of a low deep moaning repeated five or six times, ending in faintly audible sighs. At other times, he startles the forest with loud, deep-toned, solemn roars, repeated five or six times in quick succession, each increasing in loudness to the third or fourth, when his voice dies away in five or six low muffled sounds, very much resembling distant thunder. At times, and not unfrequently, a troop may be heard roaring in concert, one assuming the lead, and two, three, or four more, regularly taking up their parts, like persons singing a catch. Like our Scottish stags at the rutting-season, they roar loudest in cold frosty nights; but on no occasion are their voices to be heard in such perfection, or so intensely powerful, as when two or three strange troops of lions approach a fountain to drink at the same time. When this occurs, every member of each troop sounds a bold roar of defiance at the opposite parties, and when one roars, all roar together, and each seems to vie with his comrades in the intensity and power of his voice."

The following powerfully drawn picture, conveys a most accurate idea of the fearful banquets held in the primæval forests of Africa, and at the same time is full of interest, from the light it throws on the habits of the carnivora. Mr. Cumming had shot three rhinoceroses near a fountain, and soon after twilight had died away, he came down to the water to watch for lions. With him was his Hottentot, Kleinboy. "On reaching the water, I looked towards the carcase of the rhinoceros, and, to my astonishment, I beheld the ground alive with large creatures, as though a troop of zebras were approaching the water to drink; Kleinboy remarked to me that a troop of zebras were standing on the height: I answered 'Yes,' but I knew very well that zebras would not be capering around the carcase of a rhinoceros. I quickly arranged my blankets, pillow, and guns, in the hole, and then lay down to feast my eyes on the interesting sight before me; it was bright moonlight, as clear as I need wish. There were six large lions, about twelve or fifteen hyænas, and from twenty to thirty jackals, feasting on and around the carcases of the three rhinoceroses. The lions feasted peacefully, but the hyænas and jackals fought over every mouthful, and chased one another round and round the carcases, growling, laughing, screeching, chattering, and howling, without any intermission. The hyænas did not seem afraid of the lions, although they always gave way before them; for I observed that they followed them in the most disrespectful manner, and stood laughing, one or two on either side, when any lions came after their comrades to examine pieces of skin or bones which they were dragging away."

Lions will occasionally give chase to deer or buffaloes which have been wounded, and a very remarkable "course" of this description occurred to Mr. Oswell, an officer of the East India Company's service. This gentleman had wounded a buffalo when shooting on the banks of the river Limpopo in South Africa, and with a companion was galloping in pursuit, when suddenly three lions appeared, and, without observing the sportsmen, gave chase to the buffalo, which held on stoutly, followed by the three jolly lions, the sportsmen bringing up the rear; the lions very soon sprang on the huge buffalo, and pulled him down, when a terrific scuffle ensued; after admiring the fun for a short time, the sportsmen thought it well to interfere, and accordingly opened their fire on the lions: as these were struck by the balls, they seemed to consider them as pokes from the buffalo, and redoubled their attentions to him accordingly; at length two of the lions were killed, and the third, finding the ground too hot, made off, exceedingly puzzled at the unexpected death of his royal brothers.

According to Pliny, Hanno the Carthaginian was the first man to tame a lion:—"Primus hominum leonem manu tractare ausus, et ostendere mansuefactum, Hanno è clarissimis Pœnorum traditur."

The reign of Henry the First saw the first menagerie established in England; this monarch made at Woodstock a park, walled round with stone, seven miles in circumference, laying waste much fertile land, and destroying many villages, churches, and chapels; in the words of the old chronicler, "He appointed therein, beside great store of deer, divers strange beasts to be kept and nourished, such as were brought to him from far countries, as lions, leopards, lynxes, porpentines, and such other."

The origin of the "Lion Tower," in the Tower of London, was a present from the Emperor Frederick II. to Henry III. in 1235, of three leopards, to which he assigned quarters in that fortress. It appears that, in the reign of Edward III., one lion, one lioness, one leopard, and two "cattes lions," formed the menagerie, and were formally handed over to the custody of Robert, the son of John Bowie.

In the reigns of the first three Edwards, the allowance for each lion was sixpence a day, the wages of the keeper being three halfpence. At later periods the office of keeper of the lions was held by some person of quality about the king, with a fee of sixpence a day for himself, and the same for every lion under his charge. In 1657, there were six lions in the Tower, and not less than eleven in 1708. On the establishment of the Zoological Gardens, Regent's Park, the animals were transferred to them by William IV.

It was a curious coincidence that one of the finest litters of cubs whelped in the Tower, was born on the anniversary of Lord Howe's victory, in 1794, and the next litter was presented to the nation by the lioness, on the 20th October, 1827, the day of the battle of Navarino.

There were three very handsome young lions in the

Zoological Gardens, which were brought over from Grand Cairo by the head keeper about ten years since; he was anxious to obtain a fine female cheetah, or hunting leopard, from a person who possessed it, but he declined to part with it unless the cubs were taken also; two were mere little playthings, scarce bigger than good-sized kittens; the third, Sampson,* was larger, and had been kept chained up, which he resented exceedingly. During their voyage to England the lion cubs were great favourites, especially with the sailors, and, by way of a treat, they were now and then favoured with a fowl. The door of the poultry hutch would be opened, and out would fly a hen, cackling and rejoicing at her liberty; in a second, however, a cub would bound across the deck, make a spring, and cut short the *pæan* and the life of the poor hen together.

When newly whelped, the fur of the lion is brindled with a deep brown, especially on the legs, and there is a line of the same colour running along the back; these markings disappear during the second year. Some time since, the Society lost a lion whose history was

* This noble lion, sole survivor of the three, fell a victim to the severity of the present winter; he was apparently quite well in the evening of one of the bitterest December nights, and next morning was found dead in his den. A *post-mortem* examination disclosed the sad fact, that he had died from the intense cold, no organic disease being detected. (January, 1861.)

remarkable. Two gentlemen, brothers, were crossing a desert in Barbary, on camels, when suddenly a lioness sprang on the foremost camel; the rider of the one behind immediately fired two balls into her body with fatal effect; on examining her it was discovered that she was suckling, and two helpless young cubs were found and secured; one died, the other was reared and presented to the Gardens, where he fell a victim to the scrofulous disease, which has deprived the Society of many of their finest animals. We saw him the day before he died; he lay on his back with a deep and gaping wound in his neck, which he had considerably increased by licking with his rough tongue. It was suggested that if it could be touched with lunar caustic it might assist its healing. "Why, sir," said the keeper, "he'd be sure to *loll* the caustic, for the part is so sore, and he's so irritable, that he won't allow nothing to come nigh him, but would bite at it directly." Chloroform was suggested, but the difficulty of applying it to a lion rendered savage by pain, was the objection. Everything practicable was done, but he died the next day.

Those who visited the Gardens some years ago, may remember a remarkably fine and majestic lion, called Albert. It is not generally known that he furnished the subject for the picture by Sir E. Landseer, of the "Desert," exhibited in 1849. We were greatly amused at some of the criticisms passed on this fine portrait. "*That* a dead lion!" said one, with a knowing look; "I am sure *he* never saw a dead lion who painted *that*." Some objected to the drawing, others to the colouring; some had no patience with the background; and a few, especially wise in their generation, considered the picture as a gigantic caricature. This picture, painful to those who, like the writer, had often admired the magnificent proportions and majestic gait of this noble lion when in health, and who recognized in it a faithful delineation of nature, originated pretty much as follows. The lion was attacked with inflammation of the lungs and died, intimation was sent to the most eminent zoologist of the day, with a request to know if he wished to dissect it. Having had much experience in the anatomy of lions, he declined the opportunity, but suggested that it should be placed at the disposal of the great artist. Accordingly, about half-past five the following morning, there was a knock at Sir Edwin's bedroom door.

"Halloo! who's there?"

"Please, sir, have you ordered a lion?" was the reply.

"Ordered a *what?*"

"A lion, sir: have you ordered a lion? 'cos there's one come to the back-door, but he doesn't know whether you ordered him or not."

"Oh, very well! take him in; I'll be down directly." And the artist, rightly supposing that some friend had borne him in remembrance, but not having the most remote idea whether it was a living or a defunct lion which had thus unexpectedly paid him an early visit, hurried his toilet, and descending to his back yard, beheld the grisly monarch stretched at length upon the stones; a few minutes sufficed to arrange his materials, and so struck was he with the noble object before him, that he ceased not from his work till the picture, as exhibited, was completed.

The veterinary art must have been rather low among the Romans, if we may judge from the following ludicrous prescription for a sick lion, given us by Pliny. "The lion is never sicke but of the peevishness of his stomacke, loathing all meat; and then the way to cure him, is to ty unto him certaine shee apes, which, with their wanton mocking and making mowes at him, may move his patience, and drive him from the very indignitie of their malapert saucinesse into a fit of madnesse, and then, so soon as he hath tasted their bloud, he is perfectly wel againe; and this is the only help." *

To be licked by the tongue of a dog is a mark of affection; but such a demonstration from a lion would be productive of unpleasant consequences. The tongue in the lion and tiger tribes is covered with a thicket

* Holland's Pliny, chapter xvi.: ed. 1635.

of strong horny papillæ, the points directed backward, fitting it rather for sweeping off fragments of meat from bones, for which it is especially employed, than for gustatory enjoyment or expression of endearment. The sense of taste is very low in all the felinæ, of which an example is presented in that favourite amusement of cats, called "dressing their fur." When changing their coats the hairs are swept off in hundreds by the rough tongue without causing the slightest annoyance, whereas the presence of even a single hair in the human mouth, is notoriously unpleasant—simply from the greater perfection of the nervous influence.

One of the most remarkable things connected with the larger felinæ, is the absolute stillness with which they, huge though they be, steal upon their prey. To enable them to do this we find a special organization: the strong movable erectile hairs called whiskers, are connected with large highly sensitive nerves, so that the slightest touch is instantly felt; by the aid of these they steer their way in the darkness through the thickets without rustling a leaf. Their heavy paws also are muffled with soft fur, and their sharp claws retracted in sheaths, so that their tread is absolutely noiseless. When we consider the extraordinary acuteness of hearing and watchfulness in wild animals generally, and the great fleetness of many, we can understand why these great carnivora are furnished with such admirable means for surprising, as well as overpowering, them.

The younger Pliny, whose work on Natural History is full of information mixed up with the quaintest stories, remarks that the test of a lion's temper is his tail. "At first," says this writer, "when he entreth into his choler, he beateth the ground with his taile; when he groweth into greater heats he flappeth and jerketh his flanks and sides withall, as it were to quicken himselfe, and stir up his angry humour." Pliny, however, does not appear to have been aware of the existence of a peculiarity in the lion's tail, which was known to Didymus Alexandrinus, was subsequently denied, and rediscovered by Mr. Bennett in 1832. This is a claw at the tip of the tail, which, although not always present, undoubtedly exists in the majority of lions. Whether it has any effect in raising the "choler" of the lion it is difficult to say, but the ancient Assyrians were well acquainted with this claw, as is proved by the sculptures on the Nineveh marbles, where it is distinctly represented.

Pliny, too, picked up another story, which, although it has been ridiculed, is certainly founded on fact. "Polybius, who accompanied Scipio Æmylianus in his voiage of Africke, reporteth of them (the lions), that when they be growne aged they will prey upon a man; the reason is, because their strength will not hold out to pursue in chase any other wild beasts. Then they come about the cities and good towns of Africke, lying in wait for their prey, if any folke come abroad; and for that cause he saith that while hee was with Scipio hee saw some of them crucified and hanged up, to the end that, upon the sight of them, other lions should take example, and be skarred from doing the like mischiefe." A lion in the form of a spread eagle must have been an edifying spectacle, and it is to be hoped that the other members of the royal family profited by the example. Be that as it may, these anthropophagi still exist, and are the most dreadful scourges imaginable. The wretched Hottentots in the interior of Africa are unable to destroy them with their imperfect weapons, and night after night some poor inhabitant of the kraal is carried off, until the miserable remnant are driven to seek a precarious safety by quitting the spot, and removing perhaps to a distance of two or three hundred miles. The following account of an attack by one of these man-eaters, as they are called (for having once tasted human flesh they will eat nothing else if it can be obtained), makes the blood run cold. Mr. Cumming and his party had, unknown to them, pitched their camp in the proximity of a lion of this description; all had retired to rest, when (says Mr. C.) " suddenly the appalling and murderous voice of an angry bloodthirsty lion burst upon my ear, within a few yards of us, followed by the shrieking of the Hottentots. Again and again the murderous roar of attack was repeated. We heard John and Ruyter shriek, 'The lion! the lion!' Still for a few moments we thought he was but chasing one of the dogs round the kraal, but the next instant John Stofulus rushed into the midst of us, almost speechless with fear and terror, his eyes bursting from their sockets, and shrieked out, 'The lion! the lion! he has got Hendrick! he dragged him away from the fire beside me. I struck him with the burning brands upon his head, but he would not let go his hold. Hendrick is dead, O God! Hendrick is dead! Let us take fire and seek him.' The rest of my people rushed about shrieking and yelling as if they were mad. I was at once angry with them for their folly, and told them that if they did not stand still and keep quiet, the lion would have another of us, and that very likely there was a troop of them. I ordered the dogs, which were nearly all fast, to be made loose, and the fire to be increased as far as could be. I then shouted Hendrick's name, but all was still. I told my men that Hendrick was dead, and that a regiment of soldiers could not now help him, and hunting my dogs forward, I had everything brought within my cattle kraal, when we lighted our fire, and closed the entrance as well as we could.

"It appeared that when the unfortunate Hendrick rose to drive in the ox, the lion had watched him to his fireside, and he had scarcely lain down when the brute sprang upon him and Ruyter (for both lay under one blanket,) with his appalling, murderous roar, and roaring as he lay, grappled him with his fearful claws, and kept biting him on the breast and shoulder, all the while feeling for his neck, having got hold of which he at once dragged him away backwards round the bush into the dense shade. As the lion lay on the unfortunate man, he faintly cried, 'Help me! help me! oh God! men, help me!' after which the fearful beast got hold of his neck, and then all was still, except that his comrades heard the bones of his neck cracking between the teeth of the lion."

It is satisfactory to know that, on the following day, Mr. Cumming took revenge on the lion, whose huge grisly hide was to be seen in his collection.

The following adventure with a lion, related by the sufferer to Mr. Waterton, presents one of the most remarkable examples of courage and presence of mind under dreadful suffering on record:—

In July, 1831, two fine lions made their appearance in a jungle some twenty miles distant from the cantonment of Rajcoté, in the East Indies, where Captain Woodhouse and his two friends, Lieutenants

Delamain and Lang, were stationed. An elephant was despatched to the place in the evening on which the information arrived; and on the morrow, at the break of day, the three gentlemen set off on horseback, full of glee, and elated with the hope of a speedy engagement. On arriving at the edge of the jungle, people were ordered to ascend the neighbouring trees, that they might be able to trace the route of the lions in case they left the cover. After beating about in the jungle for some time, the hunters started the two lordly strangers. The officers fired immediately, and one of the lions fell, to rise no more. His companion broke cover, and took off across the country. The officers now pursued him on horseback as fast as the nature of the ground would allow, until they learned from the men who were stationed in the trees, that the lion had got back into the thicket. Upon this the three officers returned to the edge of the jungle, and having dismounted from their horses, they got upon the elephant, Captain Woodhouse placing himself in the hindermost seat. They now proceeded towards the heart of the jungle, in the expectation of rousing the royal fugitive a second time. They found him standing under a large bush, with his face directly towards them. The lion allowed them to approach within range of his spring, and then he made a sudden dart at the elephant, clung on his trunk with a tremendous roar, and wounded him just above the eye. While he was in the act of doing this, the two lieutenants fired at him, but without success. The elephant now shook him off, but the fierce and sudden attack on the part of the lion seemed to have thrown him into the greatest consternation. At last he became somewhat more tractable, but as he was advancing through the jungle, the lion, which had lain concealed in the high grass, made at him with redoubled fury. The officers now lost all hopes of keeping their elephant in order. He turned round abruptly, and was going away quite ungovernable, when the lion again sprang at him, seized him behind with his teeth, and hung on, until the affrighted animal managed to shake him off by incessant kicking. The lion retreated further into the thicket, Captain Woodhouse in the meantime firing a random shot at him, which proved of no avail. No exertions on the part of the officers could now force the terrified animal to face his fierce foe, and they found themselves reduced to the necessity of dismounting. Determined, however, to come to still closer quarters with the formidable king of quadrupeds, Captain Woodhouse took the desperate resolution to proceed on foot in quest of him; and after searching about for some time, he saw the lion indistinctly through the bushes, and discharged his rifle at him, but was pretty well convinced that he had not hit him, as he saw him retire with the utmost composure into the thicker parts of the brake. After some time lost in searching, the Indian gamefinder espied the lion in the cover, and pointed him out to the Captain, who fired, but unfortunately missed his mark. Having retired to reload his rifle, he was joined by Lieutenant Delamain, who, on going eight or ten paces down a sheep track, got a sight of the lion, and discharged his rifle at him. This irritated the mighty lord of the woods, and he rushed towards him, breaking through the bushes in most magnificent style. Captain Woodhouse now found himself placed in an awkward situation. He was aware that if he retraced his steps to place himself in a better position to attack, he would just get to the point from which the lieutenant had fired, and to which the lion was making. Whereupon he instantly resolved to stand still, in the hope that the lion would pass by at a distance of four yards or so without perceiving him, as the intervening cover was thick and strong. In this, however, he was unfortunately deceived, for the enraged lion saw him in passing, and flew at him with a dreadful roar. In an instant, as though by a stroke of lightning, the rifle was broken and thrown out of the Captain's hand, his left arm being at the same instant seized by the claws, and his right by the teeth of his desperate antagonist. Lieutenant Delamain now ran up, and discharged his piece full at the lion, which caused him and the Captain to come to the ground together, while Lieutenant Delamain hastened out of the jungle to reload. The lion now began to craunch the Captain's arm, but as he had the cool, determined resolution to lie perfectly still notwithstanding the dreadful pain this caused him, the lion let the arm drop out of his mouth, and quietly placed himself in a crouching position, with both of his paws upon the thigh of his fallen foe. The Captain now unthinkingly raised his hand to support his head, but no sooner had he moved it than the lion seized the lacerated arm a second time, craunched it as before, and fractured the bone still higher up. This additional *memento mori* was not lost upon Captain Woodhouse, who remained perfectly still, though bleeding and disabled under the foot of a mighty and irritated enemy. Death was close upon him, armed with every terror that could appal the heart, when, just as this world was on the point of vanishing for ever, he heard two faint reports which he thought sounded from a distance, but was totally at a loss to account for them. He afterwards learned that the reports were caused by his friend at the outside of the jungle, who had flashed off some powder in order to be quite sure that the nipples of his rifle were clean.

The two lieutenants were now hastening to his as

sistance, and he heard the welcome sound of feet approaching, but they were in a wrong direction, as the lion was between them and him. Aware that if his friends fired, the balls would hit him after they had passed through the lion's body, Captain Woodhouse quietly pronounced in a low tone, "To the other side! to the other side!" Hearing the voice, they looked in the direction from whence it proceeded, and to their horror saw their brave comrade in his utmost need. Having made a circuit, they cautiously came up on the other side, and Lieutenant Delamain fired from a distance of about a dozen yards, over the person of his prostrate friend. The lion merely quivered: his head dropped upon the ground, and in an instant he lay dead on his side close to his intended victim. Happily Captain Woodhouse, though much injured, recovered from his wounds.

In 1823, General Watson being out one morning on horseback in Bengal, armed with a double-barreled rifle, was suddenly attacked by a large lion, which bounded out from the thick jungle at the distance of only a few yards; he fired, and the lion, pierced through the heart, fell dead at his feet; but almost instantly a not less terrible opponent appeared in the lioness, who was furious at the death of her mate; but the General again fired, and wounded her so severely, that she retreated into the thicket; having loaded his rifle, he traced her to her den, and quickly gave the *coup de grâce*. In the den were found a pair of beautiful cubs, male and female, about three days old. These the General brought away with him, and fed them by means of a goat, who was prevailed on to act as their foster-mother. They were brought to England, and placed in the Tower, where both attained maturity, the lion being long known by the name of "George." He was the gentlest creature imaginable, allowing himself to be treated with the greatest familiarity by the keepers and those with whom he was acquainted: the lioness was not quite so manageable. On one occasion, when nearly full grown, she had been suffered, through inadvertence, to leave her den, when she was by no means in good temper. The under-keeper, however, alone, and armed only with a stick, had the boldness to undertake to drive her back. It was a service of no ordinary peril, for she actually made three springs at him, which he was fortunate enough to avoid; and by a bold front and determined bearing he eventually succeeded in lodging her in her place of confinement. "George" was afterwards removed to the Zoological Gardens, but did not long survive the change of quarters.

The instinct which renders the protection of the young paramount to every other consideration, is strongly evinced in the lion tribe, and of this an interesting example is narrated by Mr. Cumming. One day, when out elephant-hunting, accompanied by two hundred and fifty men, he was astonished suddenly to behold a majestic lion slowly and steadily advancing towards the party with a dignified step and undaunted bearing, the most noble and imposing that can be conceived; lashing his tail from side to side, and growling haughtily, his eyes glaring, and his teeth displayed, as he approached. The two hundred and fifty valiant men immediately took to their heels in headlong flight, and, in the confusion, four couples of dogs which they had been leading for the sportsman were allowed to escape in their couples. These instantly faced the lion, who, finding that by his bold bearing he had succeeded in putting his enemies to flight, now became solicitous for the safety of his little family with which the lioness was retreating in the background. Facing about, he followed after them with a haughty and independent step, growling fiercely at the dogs which trotted along on either side of him. Having elephants in view, the sportsman, with "heartfelt reluctance," reserved his fire, and we think that most of our readers will rejoice with us that this gallant and devoted lion was permitted to escape scot free. It would be a subject not unworthy of Landseer, this "retreat of the Leonidæ." The mother leading away the young, the noble father covering the rear, and the bold two hundred and fifty warriors in hot flight, dotting the ground in the distance. Another instance of the magnanimous conduct of the lion, is related in the case of a boer, who might well have exclaimed, "Heaven defend me from my friends!" A party of boers were out lion-hunting, when one of them, who had dismounted from his horse to get a steady shot at the lion, was dashed to the ground by him before he could regain his saddle; the lion, however, did not attempt to injure him further, but stood quietly over him lashing his tail and growling at the rest of the party, who had galloped to a distance in violent consternation. These fine fellows, instead of coming to the rescue of their comrade, opened their fire at an immense distance, the consequence of which was, that they missed the lion, and shot the man dead on the spot! The lion presently retreated, and none daring to follow him, he made good his escape.

Lichtenstein* says that the African hunters avail themselves of the circumstance that the lion does not attempt to spring upon his prey till he has measured the ground and has reached the distance of ten or twelve paces, when he lies crouching on the ground, gathering himself up for the effort. The hunters, he says, make a rule never to fire upon a lion till he lies down at this short distance, so that they can aim di-

* Travels in Southern Africa.

rectly at his head with the most perfect certainty. He adds, that, if a person has the misfortune to meet a lion, his only hope of safety is to stand perfectly still, even though the animal crouches to make his spring: that spring will not be hazarded if the man has only nerve enough to remain motionless as a statue, and look steadily in the eyes of the lion. The animal hesitates, rises, slowly retreats some steps looking earnestly about him—lies down—again retreats, till having thus by degrees quite got out of what he seems to feel as the magic circle of man's influence, he takes flight in the utmost haste.

The Field-Commandant Tjaard Van der Wolf and his brother, not far from their dwelling-house on the eastern declivity of the Snowy Mountains, followed the track of a large lion to a ravine overgrown with brushwood.

They took their stations on each side the entrance of the ravine, sending in their dogs; and presently the lion rushed towards the brother, crouched, and at the same instant received a shot from him; the shot, however, only slightly wounded him, and he made towards his assailant, who had barely time to leap on his horse and endeavour to fly. The lion was instantly after him, and sprang upon the back of his horse, who, overpowered with the burden and with fear, could no longer move. The enraged animal now stuck his claws into his victim's thigh, tearing his clothes with his teeth. The man clung with all his force to the horse, that he might not be torn off, and called to his brother for God's sake to fire, not regarding who or what he might hit. The brave Tjaard descended instantly from his horse, and taking his aim coolly, shot the lion through the head, the ball fortunately lodging in his brother's saddle without injuring either horse or rider. Less fortunate was a person of the Zwarte-Ruggens, by name Rensburg, who, with a cousin, set out on a lion-hunt. The adventure took exactly the same turn as the former, only that the lion instead of springing on the back of the horse sprang on his side, and fastened his teeth in the left arm of the rider. But how different was the conduct of this man's relative, to that of the brave Tjaard! Instead of coming to the rescue, he ran away to call some Hottentots. Rensburg, in the meantime, while the creature tore and craunched his left arm, drew a knife from his pocket with his right hand, with which he stabbed the foe in several places; and when those who were called to the rescue at length came up, they found the poor man torn from his horse and swimming in blood, his left arm and side shockingly mangled, and the dead lion with the knife still in his throat, fallen upon him. He was not then dead, but expired in a few minutes, exhausted with loss of blood.

Diederik Müller, who, next to Mr. Cumming, ranks as one of the most intrepid and successful lion-hunters in South Africa, came suddenly on a lion, who at once assumed an aspect of defiance. Diederik instantly alighted (for the boers do not seem to be in the habit of firing from a horse's back), and took deliberate aim with his rifle or *roer* at the forehead of the lion who was couched in the act of springing, but at the moment the trigger was drawn the hunter's horse started and caused him to miss his aim. The lion bounded forward, but stopped within a few paces, confronting Diederik. The man and the lion stood looking each other in the face for some minutes, and at length the lion moved backwards as if to go away. Diederik began to load his gun, the lion looked over his shoulder, gave a deep growl, and returned. Diederik stood still. The lion again moved cautiously off, and the boer proceeded to ram down his bullet. Again did the lion look back and growl angrily; and this was repeated until the animal had got off to some distance, when he took to his heels and bounded away.

Instances might be brought forward of the forbearance of lions, some of whom seem to possess a large amount of what may be termed generosity. If fairly attacked, they will fight it out; but, unless impelled by hunger, there is ample evidence to show that many are slow to destroy. Of the dangers through which the adventurous Mungo Park passed, not the least was the following.* "As we were crossing a large open plain where there were a few scattered bushes, my guide, who was a little way before me, wheeled his horse round in a moment, calling out something in the Foulah language, which I did not understand. I inquired in Mandingo what he meant. '*Wara billi billi*' (a very large lion), said he, and made signs for me to ride away. But my horse was too much fatigued, so we rode slowly past the bush from which the animal had given us the alarm. Not seeing anything myself, however, I thought my guide had been mistaken, when the Foulah suddenly put his hand to his mouth, exclaiming, '*Soubah an Allah!*' (God preserve us!), and to my great surprise, I then perceived a large red lion at a short distance from the bush, with his head crouched between his fore-paws. I expected he would instantly spring upon me, and instinctively pulled my feet from my stirrups to throw myself on the ground, that my horse might become the victim rather than myself. But it is probable the lion was not hungry, for he quietly suffered us to pass though we were fairly within his reach. My eyes were so riveted on this sovereign of the beasts, that I found it impossible to remove them until we were at a con-

* Travels in the Interior Districts of South Africa, by Mungo Park.

siderable distance." Lions are also capable of strong attachment, differing in both these respects from the tiger, who is faithless, crafty, and sanguinary. A striking illustration of this difference is afforded by a circumstance which occurred in the seventeenth century. The plague broke out at Naples with great virulence, and Sir George Davis, the English Consul there, retired to Florence. It happened that, from curiosity, he one day went to see the Grand Duke's collection of wild beasts: at the further end, in one of the dens lay a lion, which for three years had resisted every art and gentleness, continuing savage and untamable. No sooner, however, did Sir George appear in front of the den, than the lion ran to him with every mark of joy and transport. He reared himself up and licked his hand, which he had put in through the grating. The keeper, affrighted, pulled him away, begging him not to hazard his life by going so near the fiercest lion that ever entered those dens. However, nothing would satisfy Sir George but he must go into the den to him: the very instant he entered the lion threw his paws upon his shoulders, licked his face, and ran to and fro in the den, fawning and full of joy like a dog at the sight of his master. The Grand Duke, hearing of this, requested an explanation, which Sir George gave as follows. "A captain of a ship from Barbary gave me this lion when he was a young whelp. I brought him up tame, but when I thought him too large to be suffered to run about the house, I built a den for him in my courtyard. From that time he was never permitted to go loose except when I brought him within-doors to show him to my friends. When he was five years old, in his gamesome tricks he did some mischief by pawing and playing with people. Having griped a man one day a little too hard, I ordered him to be shot; upon this a friend who was at dinner with me begged him. How he came here I know not." It appeared that the gentleman who begged the lion had presented him to the Grand Duke.

With reference to the generosity of the lion, an important point turns upon the line of conduct to be pursued if a person happens to come in collision with an animal of that species, or with one of the dog tribe. With the lion, perfect quiet affords the best chance of escape. With the dog, on the contrary, resistance *à l'outrance* is necessary—it must be "death to the knife" with him—for if he overcomes his opponent, he will not cease to worry and tear so long as life exists. Some years ago, when in Lisbon, we made a short cut one night, and passing by a ruined convent which had been destroyed in the great earthquake, we suddenly came upon a pack of the savage half-wild dogs with which that city, like Constantinople, is in-

D

fested. They are the scavengers of the place, invisible during the day, but, when night falls, coming out of their lurking-places and prowling in packs, disputing with the rats the offal which the idle inhabitants throw into the streets in abundance; cowards though they are singly, they are formidable in numbers, especially to solitary passengers. There were a dozen or so in the pack the writer disturbed whilst greedily devouring some garbage, and they at once made at him. There was nothing for it but defence, so placing his back against the wall, and twisting his cloak around his left arm, a sweeping stroke with a formidable stick drove them back a few paces. Their leader was a mangy old brute with one ear and scarred in many a fight, and it was clear that the greatest danger lay in that quarter. A sharp eye was kept on him, and every time he attempted to spring he was beaten back with a blow on the nose, the others meanwhile ramping and raging in a semicircle just out of reach of the stick. This exciting amusement continued about five minutes, when fortunately a picket of soldiers turned the corner, and the curs at once fled howling. This was the first and last short cut attempted by the writer in that interesting but unclean city.

Though the lion is considerably under four feet in height, he has no difficulty in overcoming the most lofty and powerful giraffe, whose head towers above the trees, and whose skin is nearly an inch in thickness. He also, when his teeth are unbroken, generally proves a match for an old bull buffalo, which in size, strength, and fierceness, far surpasses the largest European cattle. A lion having carried off a heifer two years old, was tracked for full five hours by a party on horseback, and throughout the whole distance the carcase of the heifer was only discovered to have touched the ground twice.

The lion of South Africa is, in all respects, more formidable that the lion of India; in colour it is darker, and of greater strength; the mane, the characteristic of the male, appears about the third year; at first it is of a yellowish colour, in the prime of life nearly black, then, as he becomes aged and decrepit, it assumes a yellowish-grey or pepper-and-salt colour. The manes and coats of lions frequenting plains are richer and more bushy than those of their brethren of the forest. If the lion is thirsty, he stretches out his massive arms, lies down on his breast, and in drinking makes a loud lapping noise, pausing occasionally for breath; the tongue curls the contrary way to that of the dog during drinking.

Visitors to the Zoological Gardens cannot fail to have remarked the elegant grey *Pumas* lounging on the branches in their den, or gamboling with most graceful action. These are the representatives of the lion tribe in the New World. They have a wide geographical range, being found from the equatorial forests as far south as the cold latitudes of Tierra del Fuego. In La Plata the puma chiefly preys on deer, ostriches, and small quadrupeds, and is never dangerous to man; but in Chili it destroys horses and men. It is asserted that it always kills its prey by springing on the shoulders, and drawing back the head with one of the paws, till the neck is dislocated. Although excellent climbers, these creatures are often captured with the lasso by the guachos; at Tandeel as many as one hundred have been destroyed in three months. In Chili they are more frequently driven into trees and there shot. The puma is an exceedingly crafty animal; when pursued often doubling, and then suddenly making a powerful spring on one side, it waits till its pursuers have passed by. The flesh is eaten, and Mr. Charles Darwin gives in his Journal an amusing account of an epicurean surprise he encountered on the Rio Tapalguen. "At supper, from something which was said, I was suddenly struck with horror at thinking I was eating one of the favourite dishes of the country, namely, a half-formed calf, long before its proper time of birth. It turned out to be a puma. The meat is very white, and remarkably like veal in taste. Dr. Shaw was laughed at for stating that the flesh of lion is in great esteem, having no small affinity with veal, both in colour, taste, and flavour. Such certainly is the case with the puma. The guachos differ in opinion whether the jaguar is good eating, but are unanimous in saying that the *cat* is excellent."

Although easily tamed if captured when young, the puma is exceedingly bloodthirsty and ferocious with its prey. Of this, Colonel Hamilton Smith witnessed an extraordinary instance. A puma which had been taken, and was confined, was ordered to be shot, and was so, immediately after it had received its food. The first ball went through his body, but the only notice the animal took was by a shrill growl, redoubling his efforts to devour his food, which he continued to swallow with quantities of his own blood, till a better directed shot laid him dead.

Those who enjoyed the society of the celebrated Edmund Kean, will remember his tame puma. This fine creature was so docile and gentle, that he was often introduced to the company in the drawing-room. He was capable of strong attachment, and would lie down on his back between the feet of those he liked, and play with their garments like a huge kitten. He especially delighted in leaping and swinging about the joists of a large unoccupied room in the old College at Edinburgh. During his voyage to England this puma was on terms of intimacy with several monkeys, but a goat or a fowl was utterly irresistible; a spring, and a blow with his powerful paw, and all was over! One night, whilst in London, he made his escape into the street, but allowed himelf to be taken into custody by a watchman without even a show of resistance, trotting along by his side in the most amicable manner. After the death of this fine fellow it was discovered that a musket-ball had injured the skull, a circumstance not known during his lifetime.

When the Zoological Gardens were first established, it was considered that those animals which were natives of the Tropics required warmth, and they were, therefore, kept in close and heated rooms. The mortality was excessive, as must always be the case with animals and human beings, when densely packed in ill-ventilated dwellings; on just grounds, therefore, it was decided to try the effect of abundance of fresh air. This has answered beyond expectation, the carnivora and monkeys (among whom was the greatest mortality) having since enjoyed excellent health, and, the past winter excepted, being perfectly indifferent to cold. In their roomy dens there are large branches of trees, which, by inducing the animals to take exercise, have been fou id very beneficial. The daily allowance of food for the larger carnivora is about seven pounds of meat and bone. A good supply of water, perfect cleanliness, thorough ventilation, and careful drainage, are points specially attended to, and it would be difficult to find animals in confinement more healthy, or apparently more happy, than those which constitute the interesting collection in the Regent's Park.

CHAPTER II.

BEARS IN BRITAIN.—OUR ANCESTORS AND THEIR BEARS.—BEAR BAITING. — ZOOLOGICAL GARDENS. — THE GRIZZLY BEAR. — STRENGTH OF GRIZZLY BEAR.—ANECDOTE.—ADVENTURE.—URSINE INTERMENTS.—SHOOTING GRIZZLY BEARS.—BEARS AT THE GARDENS. — OPERATIONS FOR CARARACT. — A CHLOROFORMED BEAR.—OPERATION ON ANOTHER BEAR.—ARCTIC EXPEDITION.—A CRAFTY INTRUDER.—LAPLAND HUNTERS.—MUTUAL POLITENESS.—THE TWO FRIENDS.—AN ILL-ASSORTED COUPLE.—AN ADVENTURE.—A NARROW ESCAPE.—SWEDISH SKALLS.—AN ACCIDENT.—JAN SEVENSON.—THE BEAR AND THE CHASSEUR.—A SCALP.—AN AWFUL SITUATION.—THE GENERAL AND THE BEAR. —WOLVES AND BEARS.—A SWISS MYTH.—THE OXFORD BEAR. —TIGLATH-PILESER.—AN URSINE UNDERGRADUATE. — A BEAR IN A BED-ROOM.—AN UNEXPECTED VISITOR.—TIG ON HORSE-BACK.—A BEAR AMONG THE SAVANTS.—A BROKEN HEART.

THOSE who ramble amidst the beautiful scenery of Torquay, who gaze with admiration on the bold outlines of the Cheddar Cliffs, or survey the fertile fen district of Cambridgeshire, will find it difficult to believe that in former ages these spots were ravaged by bears surpassing in size the grizzly bear of the Rocky Mountains, or the polar bear of the arctic regions; yet the abundant remains found in Kent Hole, Torquay, and the Banwell Caves, together with those preserved in the Woodwardian Museum at Cambridge, incontestably prove that such was the case. Grand indeed was the Fauna of the British Isles in those early days! Lions—the true old British lions—as large again as the biggest African species, lurked in the ancient thickets; elephants, of nearly twice the bulk of the largest individuals that now exist in Africa or Ceylon, roamed here in herds; at least two species of rhinoceros forced their way through the primæval forests; the lakes and rivers were tenanted by hippopotami as bulky and with as great tusks as those of Africa. These statements are not the offspring of imagination, but are founded on the countless remains of these creatures which are continually being brought to light, proving from their numbers and variety of size that generation after generation had been born, and lived, and died in Great Britain.*

It is matter of history, that the brown bear was plentiful here in the time of the Romans, and was conveyed in considerable numbers to Rome to make sport in the arena. In Wales they were common beasts of chase; and in the history of the Gordons it is stated that one of that clan, so late as 1057, was directed by his sovereign to carry three bears' heads

* See 'A History of British Fossil Mammals,' by our great zoologist, Professor Owen.

on his banner, as a reward for his valour in killing a fierce bear in Scotland.

In 1252, the sheriffs of London were commanded by the king to pay fourpence a day for "our white bear in the Tower of London and his keeper;" and in the following year they were directed to provide "unum musellum et unam cathenam ferream"—*Anglicè*, a muzzle and an iron chain, to hold him when out of the water, and a long and strong rope to hold him when fishing in the Thames. This piscatorial bear must have had a pleasant time of it, as compared to many of his species, for the barbarous amusement of baiting was most popular with our ancestors. The household book of the Earl of Northumberland contains the following characteristic entry:—"Item, my Lorde usith and accustomith to gyfe yearly when hys Lordshipe is atte home to his barward, when hee comyth to my Lorde at Cristmas with his Lordshippes beests, for making his Lordschip pastyme the said xij days xxs."

In Bridgeward Without there was a district called Paris Garden; this, and the celebrated Hockley in the Hole, were in the sixteenth century the great resorts of the amateurs in bear-baiting and other cruel sports, which cast a stain upon the society of that period,—a society in a transition state but recently emerged from barbarism, and with all the tastes of a semi-barbarous people. Sunday was the grand day for these displays, until a frightful occurrence which took place in 1582. A more than usually exciting bait had been announced, and a prodigious concourse of people assembled. When the sport was at its highest, and the air rung with blasphemy, the whole of the scaffolding on which the people stood gave way, crushing many to death, and wounding many more. This was considered as a judgment of the Almighty on those Sabbath-breakers, and gave rise to a general prohibition of profane pastime on the Sabbath.

Soon after the accession of Elizabeth to the throne, she gave a splendid banquet to the French ambassadors, who were afterwards entertained with the baiting of bulls and bears (May 25, 1559). The day following, the ambassadors went by water to Paris Garden, where they patronized another performance of the same kind. Hentzer, after describing from observation a very spirited and bloody baiting, adds, " To this entertainment there often follows that of whipping a blinded bear, which is performed by five or six men, standing circularly with whips, which they exercise upon him without any mercy, as he cannot escape because of his chain. He defends himself with all his strength and skill, throwing down all that come within his reach, and are not active enough to get out of it, and tearing their whips out of their hands and breaking them." Laneham, in his account of the reception of Queen Elizabeth at Kenilworth, in 1575, gives a very graphic account of the "righte royalle pastimes." "It was a sport very pleasant to see the bear, with his pink eyes leering after his enemies' approach; the nimbleness and wait of the dog to take his advantage, and the force and experience of the bear again to avoid his assaults. If he were bitt n in one place, how he would pinch in another to get free; that if he were taken once, then by what shift with biting, with clawing, with roaring, with tossing and tumbling he would work and wind himself from them, and when he was loose, to shake his ears twice or thrice with the blood and the slaver hanging about his physiognomy."

These barbarities continued until a comparatively recent period, but are now, it is to be hoped, exploded for ever. Instead of ministering to the worst passions of mankind, the animal creation now contribute in no inconsiderable degree to the expansion of the mind and the development of the nobler feelings. Zoological collections have taken the place of the Southwark Gardens and other brutal haunts of vice, and, we are glad to say, often prove a stronger focus of attraction than the skittle-ground and its debasing society. By them laudable curiosity is awakened, and the impression, especially on the fervent and plastic minds of young people, is deep and lasting. The immense number of persons of the lower orders who visit the London Gardens prove the interest excited.* The love of natural history is inherent in the human mind; and now, for the first time, the humbler classes are enabled to see to advantage, and to appreciate the beauties of animals of whose existence they were in utter ignorance, or if known, so tinctured with the marvellous, as to cause them to be regarded mainly as objects of wonder and of dread.

California is hardly less remarkable for its bears than for its gold. The Grizzly Bear, expressively named *Ursus ferox* and *Ursus horribilis*, reigns despotic throughout those vast wilds which comprise the Rocky Mountains and the plains east of them, to latitude 61°. In size it is gigantic, often weighing 800 pounds; and we ourselves have measured a skin eight feet and a half in length. Governor Clinton received an account of one fourteen feet long, but there might have been some stretching of this skin. The claws are of great length, and cut like a chisel when the animal strikes a blow with them. The tail is so small as not to be visible; and it is a standing joke with the Indians (who with all their gravity are great wags), to desire one unacquainted with the grizzly bear to take hold

* The number of visitors to the Zoological Gardens, Regent's Park, during 1850, was 360,402; in 1851. 667,243.

of its tail. The strength of this animal may be estimated from its having been known to drag easily, to a considerable distance, the carcase of a bison, weighing upwards of a thousand pounds. Mr. Dougherty, an experienced hunter, had killed a very large bison, and having marked the spot, left the carcase for the purpose of obtaining assistance to skin and cut it up. On his return the bison had disappeared! What had become of it he could not divine; but at length, after much search, discovered it **in a deep pit,** which had been dug for it at some distance by a grizzly bear, who had carried it off and buried it during Mr. Dougherty's absence. The following incident is related by Sir John Richardson:—" A party of voyagers, who had been employed all day in tracking a canoe up the Saskatchewan, had seated themselves in the twilight by a fire, and were busy preparing their supper, when a large grizzly bear sprang over their canoe that was tilted behind them, and seizing one of the party by the shoulder, carried him off. The rest fled in terror, with the exception of a Metif, named Bourasso, who, grasping his gun, followed the bear as it was retreating leisurely with its prey. He called to his unfortunate comrade that he was afraid of hitting him if he fired at the bear; but the man entreated him to fire immediately, as the bear was squeezing him to death. On this he took a deliberate aim, and discharged his piece into the body of the bear, which instantly dropped his prey to follow Bourasso, who however escaped with difficulty, and the bear retreated to a thicket, where it is supposed to have died." The same writer mentions a bear having sprung out of a thicket, and with one blow of his paw completely scalped a man, laying bare the skull and bringing the skin down over the eyes. Assistance coming up, the bear made off, without doing him further injury; but the scalp not being replaced, the poor man lost his sight, though it is stated the eyes were uninjured.

Grizzly bears do not hug, but strike their prey with their terrific paws. We have been informed by a gentleman who has seen much of these creatures (having indeed killed five with his own hand), that when a grizzly bear sees an object, he stands up on his hind legs, and gazes at it intently for some minutes. He then, whether it be a man or a beast, goes straight on, utterly regardless of numbers, and will seize it in the midst of a regiment of soldiers. One thing only scares these creatures, and that is the *smell* of man. If in their charge they should cross a scent of this sort, they will turn and fly.

Our informant was on one occasion standing near a thicket, looking at his servant cleaning a gun. He had just dismounted, and the bridle of the thoroughbred horse was twisted round his arm. Whilst thus engaged, a very large grizzly bear rushed out of the thicket, and made at the servant, who fled. The bear then turned short upon this gentleman, in whose hand was a rifle, carrying a small ball, forty to the pound; and as the bear rose on his hind legs to make a stroke, he was fortunate enough to shoot him through the heart. Had the horse moved in the slightest at the critical moment, and jerked his master's arm, nothing could have saved him; but the noble animal stood like a rock. On another occasion, a large bear was shot mortally. The animal rushed up a steep ascent, and fell back, turning a complete somersault ere he reached the ground. The same gentleman told us two curious facts, for which he could vouch; namely, that these bears have the power of moving their claws independently. For instance, they will take up a clod of earth which excites their curiosity, and crumble it to pieces by moving their claws one on the other; and that wolves, however famished, will never touch a carcase which has been buried by a grizzly bear, though they will greedily devour all other dead bodies. The instinct of burying bodies is so strong with these bears, that instances are recorded where they have covered hunters who have fallen into their power and feigned death, with bark, grass, and leaves. If the men attempted to move, the bear would again put them down, and cover them as before, finally leaving them comparatively unhurt.

The grizzly bears have their caves, to which they retire when the cold of winter renders them torpid; and this condition is taken advantage of by the most intrepid of the hunters. Having satisfied themselves about the cave, these men prepare a candle from wax taken from the comb of wild bees, and softened by the grease of the bear. It has a large wick, and burns with a brilliant flame. Carrying this before him, with his rifle in a convenient position, the hunter enters the cave. Having reached its recesses, he fixes the candle on the ground, lights it, and the cavern is soon illuminated with a vivid light. The hunter now lies down on his face, having the candle between the back part of the cave where the bear is, and himself. In this position, with the muzzle of the rifle full in front of him, he patiently awaits his victim. Bruin is soon roused by the light, yawns and stretches himself, like a person awaking from a deep sleep. The hunter now cocks his rifle, and watches the bear turn his head and with slow and waddling steps approach the candle. This is a trying moment, as the extraordinary tenacity of life of the grizzly bear renders an unerring shot essential. The monster reaches the candle, and either lifts his paw to strike, or his nose to smell at it. The hunter steadily raises his piece; the loud report of the rifle reverberates through the cavern; and the bear falls with a heavy crash, pierced through the eye, one of the few vulnerable spots through which he can be destroyed.

The Zoological Society have at various times possessed five specimens of the grizzly bear. The first was old Martin, for many years a well-known inhabitant of the Tower Menagerie. We remember him well, as an enormous brute, quite blind from cataract, and generally to be seen standing on his hind legs, with open mouth ready to receive any titbit a compassionate visitor might bestow. Notwithstanding the length of time he was in confinement (more than twenty years), all attempts at conciliation failed, and to the last he would not permit the slightest familiarity, even from the keeper who constantly fed him. Some idea may be formed of his size, when we say that his skull (which we recently measured) exceeds in length by two inches the largest lion's skull in the Osteological Collection, although several must have belonged to magnificent animals.

After the death of Old Martin, the Society received two fine young bears from Mr. Catlin, but they soon died. Their loss, however, was replaced by three very thriving young animals from the Sierra Nevada, about eight hundred miles from San Francisco, and were brought to this country by Mr. Pacton. They were transported with infinite trouble across the Isthmus of Panama, in a box carried on men's shoulders, and were certainly the first of their race who have performed the overland journey. The price asked was £600, but they were obtained at a much less sum. An additional interest attaches to these animals from two of them having undergone the operation for cataract.

Bears are extremely subject to this disease, and of course are thereby rendered blind. Their strength and ferocity forbade anything being done for their relief, until a short time ago, when, by the aid of that wonderful agent, chloroform, it was demonstrated that they are as amenable to curative measures as the human subject.

On the 5th of November, 1850, the first operation of the sort was performed on one of these grizzly bears, which was blind in both eyes. As this detracted materially from his value, it was decided to endeavour to restore him to sight; and Mr. White Cooper having consented to operate, the proceedings were as follow:—A strong leathern collar, to which a chain was attached, was firmly buckled around the patient's neck, and the chain having been passed round one of the bars in front of the cage, two powerful men endeavoured to pull him up, in order that a sponge containing chloroform should be applied to his muzzle by Dr. Snow. The resistance offered by the bear was as surprising as unexpected. The utmost efforts of these men were unavailing; and, after a struggle of ten minutes, two others were called to their aid. By their united efforts, Master Bruin was at length brought up, and the sponge fairly tied round his muzzle. Meanwhile the cries and roarings of the patient were echoed in full chorus by his two brothers, who had been confined to the sleeping den, and who scratched and tore at the door to get to the assistance of their distressed relative. In a den on one side was the cheetah, whose leg was amputated under chloroform some months before, and who was greatly excited by the smell of the fluid and uproar. The large sloth bear in a cage on the other side, joined heartily in the chorus, and the Isabella bear just beyond, wrung her paws in an agony of woe. Leopards snarled in sympathy, and laughing hyænas swelled the chorus with their hysterical sobs. The octobasso growling of the polar bears, and roaring of the lions on the other side of the building, completed as remarkable a diapason as could well be heard.

The first evidence of the action of the chloroform on the bear, was a diminution in his struggles; first one paw dropped, then the other. The sponge was now removed from his face, the door of the den opened, and his head laid upon a plank outside. The cataracts were speedily broken up, and the bear was drawn into the cage again. For nearly five minutes he remained, as was remarked by a keeper, without knowledge, sense, or understanding, till at length one leg gave a kick, then another, and presently he attempted to stand. The essay was a failure, but he soon tried to make his way to his cage. It was Garrick, if we remember right, who affirmed that Talma was an indifferent representative of inebriation, for he was not drunk in his legs. The bear, however, acted the part to perfection, and the way in which (like Commodore Trunnion on his way to church) he tacked, during his route to his den, was ludicrous in the extreme. At length he blundered into it, and was left quiet for a time. He soon revived, and in the afternoon ate heartily. The following morning, on the door being opened, he came out, staring about him, caring nothing for the light, and began humming, as he licked his paws, with much the air of a musical amateur sitting down to a sonata on his violoncello.

A group might have been dimly seen through the fog which covered the garden, on the morning of the 15th of the same month, standing on the spot where the proceedings above narrated took place ten days previously. This group comprised Professor Owen, Mr. Yarrell, Count Nesselrode, Mr. Waterhouse, Captain Stanley, R.N., and two or three other gentlemen. They were assembled to witness a similar operation on another of the grizzly bears. The bear this time was brought out of the den, and his chain passed round the rail in front of it. Diluted chloroform was used, and the operation was rendered more difficult by the animal not being perfectly under its influence. He recovered immediately after the couching needle had been withdrawn from the second eye, and walked pretty steadily to his sleeping apartment, where he received the condolences of his brethren, rather ungraciously, it must be confessed, but his head was far from clear, and his temper ruffled. It is a singular fact that those which had been chloroformed, subsequently grew with much greater rapidity than their brother, so that there was a marked difference in size between them; but they all ultimately died from an affection resembling epilepsy, to which bears are very subject.

A recent Arctic Expedition afforded an insight into the habits and proceedings of the bears which inhabit those inhospitable regions, and the following interesting anecdotes are taken from the published report. Two bears advanced towards the exploring party commanded by Mr. M'Dougall, who thereupon shot the smaller bear through the back, paralyzing its hind quarters. On this both animals began to retreat, the wounded one being assisted by its dam in the following manner. Placing herself in such a position as to enable her cub to grasp with its forepaws her hind quarters, she trotted on with her burden faster than the party could walk, turning occasionally to watch their proceedings. At length, being wounded in the back and foot, she, maddened with rage and pain, advanced rapidly towards the party. At this critical moment Mr. M'Dougall fired and struck her in the head, from which blood flowed in large quantities. Shaking her head, and rubbing the wounded side occasionally in the snow, she now made off, leaving her young one to its fate, which was soon decided by a bullet.

Lieutenant M'Clintock says, "Shortly after pitching our tents a bear was seen approaching. The guns were prepared, men called in, and perfect silence maintained in our little camp. The animal approached rapidly from to leeward, taking advantage of every hummock to cover his advance until within seventy yards, then, putting himself in a sitting posture, he pushed forward with his hinder legs, steadying his body with his forelegs outstretched. In this manner he advanced for about ten yards further; stopped a minute or two, intently eyeing our encampment, and snuffing the air in evident doubt. Then he commenced a retrograde movement by pushing himself backward with his fore-legs, as he had previously advanced with the hinder ones. As soon as he presented his shoulder to us, Mr. Bradford and I fired, breaking a leg and otherwise wounding him severely, but it was not until he had got three hundred yards off, and received six bullets, that we succeeded in killing him."

The wooded districts of the American continent were tenanted, before civilization had made such gigantic strides, by large numbers of the well-known black bear, *Ursus Americanus*. Some years ago, black bears' skins were greatly in vogue for carriage hammercloths, etc.; and an idea of the animals destroyed, may be formed from the fact, that in 1783, 10,500 skins were imported, and the numbers steadily rose to 25,000 in 1803, since which time there has been a gradual decline. In those days, a fine skin was worth from twenty to forty guineas, but may now be obtained for five guineas.

The chase of this bear is the most solemn action of the Laplander; and the successful hunter may be known by the number of tufts of bears' hair he wears in his bonnet. When the retreat of a bear is discovered, the ablest sorcerer of the tribe beats the *runic* drum to discover the event of the chase, and on which side the animal ought to be assailed. During the attack, the hunters join in a prescribed chorus, and beg earnestly of the bear that he will do them no mischief. When dead, the body is carried home on a sledge, and the rein-deer employed to draw it is exempt from labour during the remainder of the year.

A new hut is constructed for the express purpose of cooking the flesh, and the huntsmen, joined by their wives, sing again their songs of joy and of gratitude to the animal, for permitting them to return in safety. They never presume to speak of the bear with levity, but always allude to him with profound respect, as "the old man in the fur cloak." The Indians, too, treat him with much deference. An old Indian, named Keskarrah, was seated at the door of his tent, by a small stream, not far from Fort Enterprise, when a large bear came to the opposite bank, and remained for some time apparently surveying him. Keskarrah, considering himself to be in great danger, and having no one to assist him but his aged wife, made a solemn speech, to the following effect:—"Oh, bear! I never did you any harm; I have always had the highest respect for you and your relations, and never killed any of them except through necessity. Pray, go away, good bear, and let me alone, and I promise not to molest you." The bear (probably regarding the old gentleman as rather a tough morsel) walked off, and the old man, fancying that he owed his safety to his eloquence, favoured Sir John Richardson with his speech at length. The bear in question, however, was of a different species to, and more sanguinary than, the black bear, so that the escape of the old couple was regarded as remarkable.

The *Ursus Americanus* almost invariably hybernates; and about a thousand skins have been annually imported by the Hudson's Bay Company, from these black bears destroyed in their winter retreats. A spot under a fallen tree is selected for its den, and having scratched away a portion of the soil, the bear retires thither at the commencement of a snow-storm, and the snow soon furnishes a close warm covering. When taken young, these bears are easily tamed: and the following incident occurred to a gentleman of our acquaintance. A fine young bear had been brought up by him with an antelope of the elegant species called *Furcifer*, the two feeding out of the same dish, and being often seen eating the same cabbage. He was in the habit of taking these pets out with him, leading the bear by a string. On one occasion he was thus proceeding, a friend leading the antelope, when a large fierce dog flew at the latter. The gentleman, embarrassed by his charge, called out for assistance to my informant, who ran hastily up, and in doing so accidentally let the bear loose. He seemed to be perfectly aware that his little companion was in difficulty, and rushing forward, knocked the dog over and over with a blow of his paw, and sent him off howling. The same bear would also play for hours with a bison calf, and when tired with his romps, jumped into a tub to rest; having recovered, he would spring out and resume his gambols with his boisterous playfellow, who seemed to rejoice when the bear was out of breath, and could be taken at a disadvantage, at which time he was sure to be pressed doubly hard. There was a fine bear of this description in the old Tower Menagerie, who long shared his den with a hyæna, with whom he was on good terms except at meal-times, when they would quarrel in a very ludicrous manner, for a piece of beef, or whatever else might happen to form a bone of contention between them. The hyæna, though by far the smaller, was generally master, and the bear would moan most piteously in a tone resembling the bleating of a sheep, while the hyæna quietly consumed the remainder of the dinner.

The following is an account of an adventure which occurred to Frank Forester, in America. A large bear was traced to a cavern in the Round Mountain, and every effort made for three days without success to smoke or burn him out. At length a bold hunter, familiar with the spot, volunteered to beard the bear in his den. The well-like aperture, which alone could be seen from without, descended for about eight feet, then turned sharp off at right angles, running nearly horizontally for about six feet, beyond which it opened into a small circular chamber, where the bear had taken up his quarters. The man determined to descend, to worm himself, feet forward, on his back, and to shoot at the eyes of the bear, as they would be visible in the dark. Two narrow laths of pine-wood were accordingly procured, and pierced with holes, in which candles were placed and lighted. A rope was next made fast about his chest, a butcher's knife disposed in readiness for his grasp, and his musket loaded with two good ounce bullets, well wrapped in greased buckskin. Gradually he disappeared, thrusting the lights before him with his feet, and holding the musket ready cocked in his hand. A few anxious moments—a low stifled growl was heard—then a loud, bellowing, crashing report, followed by a wild and fearful howl, half anguish, half furious rage. The men above wildly and eagerly hauled up the rope, and the sturdy hunter was whirled into the air uninjured, and retaining in his grasp his good weapon; while the fierce brute rushed tearing after him even to the cavern's mouth. As soon as the man had entered the small chamber, he perceived the glaring eyeballs of the bear, had taken steady aim at them, and had, he believed, lodged his bullets fairly. Painful moanings were soon heard from within, and then all was still! Again the bold man determined to seek the monster; again he vanished, and his musket shot roared from the recesses of the rock. Up he was whirled, but this time the bear, streaming with gore, and furious with pain, rushed after him, and with a mighty bound cleared the confines of the cavern. A hasty and harmless volley was fired, whilst the bear glared round as if undecided upon which of the group to wreak his vengeance. Tom, the hunter, coolly raised his piece, but snap! no spark followed the blow of the hammer! With a curse Tom threw down the musket, and drawing his knife, rushed forward to encounter the bear single-handed. What would have been his fate, had the bear folded him in his deadly hug, we may be pretty sure; but ere this could happen, the four bullets did their work, and he fell: a convulsive shudder passed through his frame, and all was still. Six hundred and odd pounds did he weigh, and great were the rejoicings at his death.

The wild pine-forests of Scandinavia yet contain bears in considerable numbers. The general colour of these European bears is dark brown, and, to a great degree, they are vegetable feeders, although exceedingly fond of ants and honey. Their favourite food is berries and succulent plants; and in autumn, when the berries are ripe, they become exceedingly fat. Towards the end of November the bear retires to his den, and passes the winter months in profound repose. About the middle of April he leaves his den, and roams about the forest ravenous for food. These bears attain a large size, often weighing above four hundred pounds;

and an instance is on record of one having weighed nearly seven hundred and fifty pounds. The best information relative to the habits and pursuits of these Scandinavian bears is to be found in Mr. Lloyd's 'Field Sports of the North of Europe,' from which entertaining work we shall draw largely.

When a district in Sweden is infested with bears, public notice is given from the pulpit during divine service, that a skäll or battue is to take place, and specifying the number of people required, the time and place of rendezvous, and other particulars. Sometimes as many as fifteen hundred men are employed, and these are regularly organized in parties and divisions. They then extend themselves in such a manner that a cordon is formed, embracing a large district, and all simultaneously move forward. By this means the wild animals are gradually driven into a limited space, and destroyed as circumstances admit. These skälls are always highly exciting, and it not unfrequently happens that accidents arise, from the bears turning upon and attacking their pursuers. A bear which had been badly wounded, and was hard pressed, rushed upon a peasant whose gun had missed fire, and seized him by the shoulder with his forepaws. The peasant, for his part, grasped the bear's ears. Twice did they fall, and twice get up, without loosening their holds, during which time the bear had bitten through the sinews of both arms, from the wrists upwards, and was approaching the exhausted peasant's throat, when Mr. Falk, "öfwer jäg mästare," or head ranger of the Wermeland forests, arrived, and with one shot ended the fearful conflict.

Jan Svenson was a Dalecarlian hunter of great repute, having been accessory to the death of sixty or seventy bears, most of which he had himself killed. On one occasion he had the following desperate encounter:—Having, with several other peasants, surrounded a very large bear, he advanced with his dog to rouse him from his lair; the dog dashed towards the bear, who was immediately after fired at and wounded by one of the peasants. This man was prostrated by the infuriated animal, and severely lacerated. The beast now retraced his steps, and came full on Jan Svenson, a shot from whose rifle knocked him over. Svenson, thinking the bear was killed, coolly commenced reloading his rifle. He had only poured in the powder, when the bear sprang up and seized him by the arm. The dog, seeing the jeopardy in which his master was placed, gallantly fixed on the bear's hind quarters. To get rid of this annoyance, the bear threw himself on his back, making with one paw a blow at the dog, with the other holding Svenson fast in his embraces. This he repeated three several times, handling the man as a cat would a mouse, and in the intervals he was biting him in different parts of the body, or standing still as if stupefied. In this dreadful situation Svenson remained nearly half an hour; and during all this time the noble dog never ceased for a moment his attacks on the bear. At last the brute quitted his hold, and moving slowly to a small tree at a few paces' distance, seized it with his teeth; he was in his last agonies, and presently fell dead to the ground. On this occasion Svenson was wounded in thirty-one different places, principally in the arms and legs. This forest monster had, in the early part of the winter, mortally wounded another man who was pursuing him, and from his great size was an object of general dread.

Lieutenant Oldenburg, when in Torp in Norrland, saw a chasseur brought down from the forest, who had been desperately mangled by a bear. The man was some distance in advance of his party, and wounded the animal with a ball. The bear immediately turned on him; they grappled, and both soon came to the ground. Here a most desperate struggle took place, which lasted a considerable time, sometimes the man, who was a powerful fellow, being uppermost, at other times the bear. At length, exhausted with fatigue and loss of blood, the chasseur gave up the contest, and turning on his face in the snow, pretended to be dead. Bruin, on this, quietly seated himself on his body, where he remained for near half an hour. At length the other chasseurs came up, and relieved their comrade by shooting the bear through the heart. Though terribly lacerated, the man eventually recovered.

Captain Eurenius related to Mr. Lloyd an incident which he witnessed in Wenersborg, in 1790:—A bear-hunt or skäll was in progress, and an old soldier placed himself in a situation where he thought the bear would pass. He was right in his conjecture, for the animal soon made his appearance, and charged directly at him. He levelled his musket, but the piece missed fire. The bear was now close, and he attempted to drive the muzzle of the gun down the animal's throat. This attack the bear parried like a fencing-master, wrested the gun from the man, and quickly laid him prostrate. Had he been prudent all might have ended well, for the bear, after smelling, fancied him dead, and left him almost unhurt. The animal then began to handle the musket, and knock it about with his paws. The soldier seeing this, could not resist stretching out his hand, and laying hold of the muzzle, the bear having the stock firmly in his grasp. Finding his antagonist alive, the bear seized the back of his head with his teeth, and tore off the whole of his scalp, from the nape of the neck upwards, so that it merely hung to the forehead by a strip of skin. Great as was his agony, the poor fellow kept quiet, and the bear laid himself along his body. Whilst this was going forward, Captain Eurenius and others approached the spot, and on coming within sixteen paces, beheld the bear licking the blood from the bare skull, and eyeing the people, who were afraid to fire lest they should injure their comrade. Captain Eurenius asserted, that in this position the soldier and bear remained for a considerable time, until at last the latter quitted his victim, and slowly began to retire, when a tremendous fire being opened, he fell dead. On hearing the shots, the wretched sufferer jumped up, his scalp hanging over his face, so as to completely blind him. Throwing it back with his hand, he ran towards his comrades like a madman, frantically exclaiming, "The bear! the bear!" The scalp was separated, and the Captain described it as exactly resembling a peruke. In one respect the catastrophe was fortunate for the poor soldier; it was in the old days of pipeclay and pomatum, and every one in the army was obliged to wear his hair of a certain form, and this man being, for satisfactory reasons, unable to comply with the regulation, and a tow wig not being admissible, he immediately received his discharge.

Mr. Paget, in his interesting work on Hungary and Transylvania, gives the following amusing account of an adventure with a bear, which took place in the neighbourhood of Kronstadt:—"General V——, the Austrian commander of the forces of the district, had come to Kronstadt to inspect the troops, and had been invited by our friend (Herr von L——), to join him in a bear-hunt. Now the General, though more accustomed to drilling than hunting, accepted the invitation, and appeared in due time in a cocked hat and long grey coat, the uniform of an Austrian general. When they had taken up their places, the General, with half-a-dozen rifles arranged before him, paid such devoted attention to a bottle of spirits he had brought with him, that he quite forgot the object of his coming. At last, however, a huge bear burst suddenly from the covert of the pine forest, directly in front of him: at that moment the bottle was raised so high that it quite obscured the General's vision, and he did not perceive the intruder till he was close upon him. Down went the bottle—up jumped the astonished soldier, and forgetful of his guns, off he started, with the bear clutching at the tails of his great coat as he ran away. What strange confusion of ideas was muddling the General's intellect at the moment, it is difficult to say: but I suspect he had some notion that the attack was an act of insubordination on the part of Bruin, for he called out most lustily as he ran along, 'Back, rascal! back! I am a General!' Luckily a poor Wallach peasant had more respect for the epaulets than the bear, and throwing himself in the way with nothing but a spear for his defence, he kept the enemy at bay till our friend and the jägers came up and finished the contest with their rifles."

A curious circumstance is related by Mr. Lloyd, showing the boldness of wolves when pressed by hunger. A party were in chase of a bear, who was tracked by a dog. They were some distance behind the bear, when a drove of five wolves attacked and devoured the dog. Their appetites being thus whetted, they forthwith made after the bear, and coming up with him, a severe conflict ensued, as was apparent from the quantity of hair, both of the bear and wolves, that was scattered about the spot. Bruin was victorious, but was killed a few days afterwards by the hunters. The wolves, however, had made so free with his fur, that his skin was of little value. On another occasion, a drove of wolves attacked a bear, who, posting himself with his back against a tree, defended himself for some time with success; but at length his opponents contrived to get under the tree, and wounded him desperately in the flank. Just then some men coming up the wolves retreated, and the wounded bear became an easy prey.

It occasionally happens that cattle are attacked by bears, but the latter are not always victorious. A powerful bull was charged in the forest by a bear, when, striking his horns into his assailant, he pinned him to a tree. In this situation they were both found dead,—the bull from starvation, the bear from wounds. So says the author above quoted.

The hybernation of bears gives rise to a curious confusion of cause and effect in the minds of the Swiss peasantry. They believe that bears which have passed the winter in the mountain caverns, always come out to reconnoitre on the 2nd of February; and that they, if the weather be then cold and winterly, return, like the dove to the ark, for another fortnight; at the end of which time they find the season sufficiently advanced to enable them to quit their quarters without inconvenience; but that, if the weather be fine and warm on the 2nd, they sally forth, thinking the winter past. But on the cold returning after sunset, they discover their mistake, and return in a most sulky state of mind, without making a second attempt until after the expiration of six weeks, during which time man is doomed to suffer all the inclemencies consequent on their want of urbanity. Thus, instead of attributing the retirement of the bears to the effects of the cold, the myth makes the cold to depend on the seclusion of the bears.

The fat of bears has, from time immemorial, enjoyed a high reputation for promoting the growth of hair; but not a thousandth part of the bear's grease sold in shops comes from the animal whose name it carries. In Scandinavia, the only part used for the hair is the fat found about the intestines. The great bulk of the fat, which in a large bear may weigh from sixty to eighty pounds, is used for culinary purposes. Bears' hams, when smoked, are great delicacies, as are also the paws; and the flesh of bears is not inferior to excellent beef.

On a certain memorable day, in 1847, a large hamper reached Oxford, per Great Western Railway, and was in due time delivered according to its direction at Christchurch, consigned to Francis Buckland, Esq., a gentleman well known in the University for his fondness for natural history. He opened the hamper, and the moment the lid was removed, out jumped a creature about the size of an English sheep-dog, covered with long shaggy hair, of a brownish colour. This was a young bear, born on Mount Lebanon, in Syria, a few months before, who had now arrived to receive his education at our learned University. The moment that he was released from his irksome attitude in the hamper, he made the most of his liberty, and the door of the room being open, he rushed off down the cloisters. Service was going on in the chapel, and, attracted by the pealing organ, or some other motive, he made at once for the chapel. Just as he arrived at the door, the stout verger happened to come thither from within, and the moment he saw the impish looking creature that was running into his domain, he made a tremendous flourish with his silver wand, and darting into the chapel, ensconced himself in a tall pew, the door of which he bolted. Tiglath Pileser (as the bear was called) being scared by the wand, turned from the chapel, and scampered frantically about the large quadrangle, putting to flight the numerous parties of dogs, who in those days made that spot their afternoon rendezvous. After a sharp chase, a gown was thrown over Tig, and he was with difficulty secured. During the struggle, he got one of the fingers of his new master into his mouth, and—did he bite it off? No, poor thing! but began vigorously sucking it, with that peculiar mumbling noise for which bears are remarkable. Thus was he led back to Mr. Buckland's rooms, walking all the way on his hind legs, and sucking the finger with all his might. A collar was put round his neck, and Tig became a prisoner. His good-nature and amusing tricks soon made him a prime favourite with the under-graduates; a cap and gown were made, attired in which (to the great scandal of the *dons*) he accompanied his master to breakfasts and wine parties, where he contributed greatly to the amusement of the company, and partook of good things, his favourite viands being muffins and ices. He was in general of an amiable disposition, but subject to fits of rage, during which his violence was extreme; but a kind word, and a finger to suck, soon brought him round. He was most impatient of solitude, and would cry for hours when left alone, particularly if it was dark. It was this unfortunate propensity which brought him into especial disfavour with the late Dean of Christchurch, whose Greek quantities and hours of rest were sadly disturbed by Tig's lamentations.

On one occasion he was kept in college till after the gates had been shut, and there was no possibility of getting him out without the porter seeing him, when there would have been a fine of ten shillings to pay, the next morning; for during this term an edict had gone forth against dogs, and the authorities, not being learned in zoology, could not be persuaded that a bear was not a dog. Tig was, therefore, tied up in a courtyard near his master's rooms, but that gentleman was soon brought out by his piteous cries, and could not pacify him in any other way than by bringing him into his rooms; and at bed-time Tig was chained to the post at the bottom of the bed, where he remained quiet till daylight, and then shuffling on to the bed, awoke his master by licking his face: he took no notice, and presently Tig deliberately put his hind legs under the blankets and covered himself up; there he remained till chapel time, when his master left him, and on his return found that the young gentleman had been amusing himself during his solitude by overturning everything he could get at in the room, and, apparently, had had a quarrel and fight with the looking-glass, which was broken to pieces and the wood-work bitten all over. The perpetrator of all this havoc sat on the bed, looking exceedingly innocent, but rocking backwards and forwards as if conscious of guilt and doubtful of the consequences.

Near to Tig's house there was a little monkey tied to a tree, and Jacko's great amusement was to make grimaces at Tig; and when the latter composed himself to sleep in the warm sunshine, Jacko would cautiously descend from the tree, and twisting his fingers in Tig's long hair, would give him a sharp pull and in a moment be up the tree again, chattering and clattering his chain. Tig's anger was most amusing: he would run backwards and forwards on his hind legs, sucking his paws, and with his eyes fixed on Jacko, uttering all sorts of threats and imprecations, to the great delight of the monkey. He would then again endeavour to take a nap, only to be again disturbed by his little tormentor. However, these two animals established a truce, became excellent friends, and would sit for half an hour together confronting each other, apparently holding a conversation. At the commencement of the long vacation, Tig, with the other members of the University, retired into the country, and was daily taken out for a walk round the village, to the great astonishment of the bumpkins. There was a little shop, kept by an old dame who sold whipcord, sugar-candy, and other matters, and here, on one occasion, Tig was treated to sugar-candy. Soon afterwards he got loose, and at once made off for the shop, into which he burst, to the unutterable terror of the spectacled and high-capped old lady, who was knitting stockings behind the counter;—the moment she saw his shaggy head and heard the appalling clatter of his chain, she rushed upstairs in a delirium of terror. When assistance arrived, the offender was discovered seated on the counter, helping himself most liberally to brown sugar; and it was with some difficulty, and after much resistance, that he was dragged away.

Mr. Buckland had made a promise that Tig should pay a visit to a village about six miles distant, and determined that he should proceed thither on horseback. As the horse shied whenever the bear came near him, there was some difficulty in getting him mounted; but at last his master managed to pull him up by the chain while the horse was held quiet. Tig at first took up his position in front, but soon walked round and stood up on his hind-legs, resting his forepaws on his master's shoulders. To him this was exceedingly pleasant, but not so to the horse, who, not being accustomed to carry two, and feeling Tig's claws, kicked and plunged to rid himself of the extra passenger. Tig held on like grim death, and stuck in his claws most successfully; for in spite of all the efforts of the horse he was not thrown. In this way the journey was performed, the countryfolks opening their eyes at the apparition.

This reminds us of an anecdote mentioned by Mr. Lloyd. A peasant had reared a bear, which became so tame that he used occasionally to cause him to stand at the back of his sledge when on a journey; but the bear kept so good a balance that it was next to impossible to upset him. One day, however, the peasant amused himself by driving over the very worst ground he could find, with the intention, if possible, of throwing Bruin off his equilibrium. This went on for some time, till the animal become so irritated that he gave his master, who was in front of him, a tremendous thump on the shoulder with his paw, which frightened the man so much, that he caused the bear to be killed immediately; this, as he richly deserved the thump, was a shabby retaliation.

When term recommenced, Tiglath Pileser returned to the University, much altered in appearance, for being of the family of silver bears of Syria, his coat had become almost white; he was much bigger and stronger, and his teeth had made their appearance, so that he was rather more difficult to manage; the only way to restrain him, when in a rage, was to hold him by the ears; but on one occasion, having lost his temper, he tore his cap and gown to pieces. About this time the British Association paid a visit to Oxford, and Tig was an object of much interest. The writer was present on several occasions when he was introduced to breakfast parties of eminent savants, and much amusement was created by his tricks, albeit they were a little rough. In more than one instance he made sad havoc with book-muslins and other fragile articles of female attire; on the whole, however, he conducted himself with great propriety, especially at an evening meeting at Dr. Daubeny's, where he was much noticed, to his evident pleasure.

However, the authorities at Christchurch, not being zoologists, had peculiar notions respecting bears; and at length, after numerous threats and pecuniary penalties, the fatal day arrived, and Tig's master was informed that either " he or the bear must leave Oxford the next morning." There was no resisting this, and poor dear Tig was accordingly put into a box—a much larger one than that in which he had arrived—and sent off to the Zoological Gardens, Regent's Park. Here he was placed in a comfortable den by himself; but, alas! he missed the society to which he had been accustomed, the excitement of a college life, and the numerous charms by which the University was endeared to him: he refused his food: ran perpetually up and down his den in the vain hope to escape, and was one morning found dead, a victim to a broken heart.

CHAPTER III.

ADAPTATION OF COLOURS.—HIGHLAND TARTANS.—VARIETIES.—TEMPERS.—TAME LEOPARDS.—THE MILLINERS' FRIENDS.—FAIR RETALIATION.—ANECDOTE OF A PANTHER.—SAÏ AND HIS KEEPER.—AN ALARM.—AFFECTION OF A PANTHER.—SAÏ AND THE ORANG.—SHORT ALLOWANCE.—MR. ORPEN'S ENCOUNTER.—TENACITY OF LIFE.—A CONFLICT.—MAJOR DENHAM.—LURKING PANTHERS.—TRAPPING A LEOPARD.—SOUTH AFRICAN LEOPARD.—PLEASANT SURPRISE.—TREE TIGERS.—EAGER SPORTSMEN.—THE TREE TIGER AND ARTILLERYMEN.—A PRACTICAL JOKE.—SCENERY OF THE ORINOCO.—AN AGREEABLE NEIGHBOURHOOD.—JAGUAR AND VULTURES.—AL FRESCO TROUBLES.—A BOLD THIEF.—JAGUAR AND TURTLE.—MODE OF ATTACK.—SACRILEGE.—FAVOURITE TREES.—LEOPARDS IN TREES.—THE GUACHO AND THE LEOPARD. — NARROW ESCAPE.—POISONED ARROWS. — A ROUGH PLAYFELLOW.—ADVANTAGE OF POLITENESS.—VALUE OF SKINS.

THERE is no class of animals which combines, in such a marked degree, beauty of form, with a wily and savage nature, as that to which the Leopard tribe belongs. The unusual pliability of the spine and joints with which they are endowed, imparts agility, elasticity, and elegance to their movements, whilst the happy proportions of their limbs give grace to every attitude. Their skins, beautifully sleek, yellow above and white beneath, are marked with spots of brilliant black disposed in patterns according to the species; nor are these spots for ornament alone; as was remarked by one of the ablest of the writers in the 'Quarterly,' the different and characteristic markings of the larger feline animals bear a direct relation to the circumstances under which they carry on their predatory pursuits. The tawny colour of the lion harmonizes with the parched grass or yellow sand, along which he steals towards, or on which he lies in wait to spring upon, a passing prey; and a like relation to the place in which other large feline animals carry on their predatory pursuits, may be traced to their different and characteristic markings. The royal tiger, for instance, which stalks or lurks in the jungle of richly-wooded India, is less likely to be discerned as he glides along the straight stems of the underwood, by having the tawny ground-colour of his coat variegated by dark vertical stripes, than if it were uniform like the lion's. The leopard and panther again, which await the approach of their prey crouching on the outstretched branch of some tree, derive a similar advantage, by having the tawny ground-colour broken by dark spots like the leaves around them; but amidst all this variety, in which may be traced the principles of adaptation to special ends, there is a certain unity of plan, the differences not being established from the beginning. Thus the young lion is spotted during his first year with dark spots on its lighter brown, and transitorily shows the livery that is most common in the genus. It is singular that man has, in a semi-barbarous state, recognized the same principle as that which constitutes these differences, and applied it to the same purpose. It is well known that the *setts*, or patterns, of several of the Highland tartans were originally composed with special reference to concealment among the heather. And with the Highlanders, perhaps, the hint was taken from the ptarmigans and hares of their own native mountains, which change their colours with the season, donning a snow-white vest when the ground on which they tread bears the garb of winter, and resuming their garments of greyish brown when the summer's sun has restored to the rocks their natural tints.

There are three species sufficiently resembling each other in size and general appearance, to be confounded by persons unacquainted with their characteristics, namely, the leopard, the panther, and the jaguar. The precise distinction between the first two is still an open question, although the best authorities agree in considering that they are distinct animals; still confusion exists. An eminent dealer in furs informed us that in the trade, panther skins were looked upon as being larger than leopards', and the spots more irregular, but the specimens produced were clearly jaguar skins, which made the matter more complicated.

The panther, *Felis pardus*, is believed to be an inhabitant of a great portion of Africa, the warmer parts of Asia, and the islands of the Indian Archipelago; while the leopard, *Felis leopardus*, is thought to be confined to Africa. The jaguar, *Felis onca*, is the scourge of South America, from Paraguay almost to the Isthmus of Darien, and is altogether a larger and more powerful animal than either of the others. Though presenting much resemblance, there are points of distinction by which the individual may be at once recognized. The jaguar is larger, sturdier, and altogether more thickset than the leopard, whose limbs are the *beau idéal* of symmetry and grace. The leopard is marked with numerous spots, arranged in small, irregular circles on the sides, the ridge of the back, the head, neck, and limbs being simply spotted, without order. The jaguar is also marked with black spots, but the circles formed by them are much larger, and in almost all a central spot exists, the whole bearing a rude resemblance to a rose; along the back, the spots are so narrow and elongated, as to resemble stripes. The tail of the jaguar is also considerably shorter than that of the leopard, which is nearly as long as the whole body.

Leopards and panthers, if taken quite young, and treated with kindness, are capable of being thoroughly tamed; the poet Cowper describes the great difference in the dispositions of his three celebrated hares; so it is with other wild animals, and leopards among the rest, some returning kindness with the utmost affection, others being rugged and untamable from the first. Of those brought to this country, the characters are much influenced by the treatment they have experienced on board ship; in some cases, they have been made pets by the sailors, and are as tractable as domestic cats; but when they have been teased and subjected to ill-treatment during the voyage, it is found very difficult to render them sociable; there are now (1852) six young leopards in one den at the Zoological Gardens: of these, five are about the same age, and grew up as one family; the sixth was added some time after, and being looked upon as an intruder, was quite sent to Coventry, and even ill-treated by the others; this he has never forgotten. When the keeper comes to the den, he courts his caresses and shows the greatest pleasure; but if any of his companions advance to share them with him, he growls and spits, and shows the utmost jealousy and displeasure.

In the same collection there is a remarkably fine, full-grown leopard, presented by her Majesty, which is as tame as any creature can be; mutton is his favourite food, but the keeper will sometimes place a piece of beef in the den; the leopard smells it, turns it over with an air of contempt, and coming forward, peers round behind the keeper's back to see if he has not (as is generally the case) his favourite food concealed. If given to him, he lays it down, and will readily leave it at the keeper's call to come and be patted, and whilst caressed he purrs, and shows the greatest pleasure.

There were a pair of leopards in the Tower before the collection was broken up, which illustrated well the difference in disposition; the male, a noble animal, continued to the last as sullen and savage as on the day of his arrival. Every kindness was lavished upon him by the keepers, but he received all their overtures with such a sulky and morose return, that nothing could be made of his unreclaimable and unmanageable disposition. The female, which was the older of the two, on the contrary, was as gentle and affectionate as the other was savage, enjoying to be patted and caressed by the keeper, and fondly licking his hands; one failing, however, she had, which brought affliction to the soul of many a beau and lady fair; it was an extraordinary predilection for the destruction of hats, muffs, bonnets, umbrellas, and parasols, and indeed, articles of dress generally, seizing them with the greatest quickness, and tearing them into pieces, almost before the astonished victim was aware of the loss; to so great an extent did she carry this peculiar taste, that Mr. Cops, the superintendent, used to say that she had made prey of as many of these articles as there were days in the year. Animals in menageries are sometimes great enemies to the milliner's art; giraffes have been known to filch the flowers adorning a bonnet, and we once saw a lady miserably oppressed by monkeys. She was very decidedly of "a certain age," but dressed in the extreme of juvenility, with flowers and ribbons of all the colours of the rainbow. Her complexion was delicately heightened with rouge, and the loveliest tresses played about her cheeks. As she languidly sauntered through the former monkey-house at the gardens, playfully poking the animals with her parasol, one seized it so vigorously, that she was drawn close to the den; in the twinkling of an eye, a dozen little paws were protruded, off went bonnet, curls and all, leaving a deplorably grey head, whilst others seized her reticule and her dress, pulling it in a very unpleasant manner. The handiwork of M. Vouillon was of course a wreck, and the contents of the reticule, her purse, gloves, and delicately-scented handkerchief, were with difficulty recovered from out of the cheek-pouch of a baboon.

On another occasion we saw the elephant, that fine old fellow who died some years ago, administer summary punishment to a weak-minded fop, who kept offering him cakes, and on his putting out his trunk, withdrawing them, and giving him a rap with his cane instead. One of the keepers warned him, but he laughed, and after he had teased the animal to his heart's content, walked away. After a time he was strolling by the spot again, intensely satisfied with himself, his glass stuck in his eye and smiling blandly in the face of a young lady who was evidently offended at his impudence, when the elephant, who was rocking backwards and forwards, suddenly threw out his trunk and seized our friend by the coat-tails; the cloth gave way, and the whole back of the coat was torn out, leaving nothing but the collar, sleeves, and front. As may be supposed, this was a damper; indeed, we never saw a man look so small, as he shuffled away amidst the titters of the company, who enjoyed his just reward.

That very agreeable writer, Mrs. Lee, formerly Mrs. Bowdich, has related, in the first volume of the 'Magazine of Natural History,' a most interesting account of a tame panther which was in her possession several months. He and another were found very young in the forest, apparently deserted by their mother; they were taken to the King of Ashantee, in whose palace they lived some weeks, when our hero, being much larger than his brother, suffocated him in a fit of romping, and was then sent to Mr. Hutchinson, the resident left by Mr. Bowdich at Coomassie, by whom he was tamed. When eating was going on, he would sit by his master's side, and receive his share with gentleness. Once or twice he purloined a fowl, but easily gave it up on being allowed a portion of something else; but on one occasion, when a silly servant tried to pull his food from him, he tore a piece of flesh from the offender's leg, but owed him no ill-will afterwards. One morning he broke the cord by which he was confined, and the castle-gates being shut, a chase commenced; but after leading his pursuers several times round the ramparts, and knocking over a few children by bouncing against them, he suffered himself to be caught, and led quietly back to his quarters, under one of the guns of the fortress. By degrees all fear of him subsided, and he was set at liberty, a boy being appointed to prevent his intruding into the apartments of the officers. His keeper, however, like a true negro, generally passed his watch in sleeping, and Saï, as the panther was called, roamed at large. On one occasion he found his servant sitting on the step of the door, upright, but fast asleep, when he lifted his paw, gave him a pat on the side of the head which laid him flat, and then stood wagging his tail as if enjoying the joke. He became exceedingly attached to the governor, and followed him everywhere like a dog. His favourite station was at a window in the sitting-room, which overlooked the whole town; there, standing on his hind legs, his fore-paws resting on the ledge of the window, and his chin laid between them, he amused himself with watching all that was going on. The children were also fond of this scene; and once, finding Saï's presence an encumbrance, they united their efforts, and pulled him down by the tail. He one day missed the governor, and wandered with a dejected look to various parts of the fortress in search of him; whilst absent on this errand, the governor returned to his private room and seated himself at a table to write; presently he heard a heavy step coming up the stairs, and raising his eyes to the open door, beheld Saï. At that moment he gave himself up for lost, for Saï immediately sprang from the door on to his neck: instead, however, of devouring him, he laid his head close to the governor's, rubbed his cheek upon his shoulder, wagged his tail, and tried to evince his happiness. Occasionally, however, the panther caused a little alarm to the other inmates of the castle, and on one occasion the woman whose duty it was to sweep the floors, was made ill by her fright; she was sweeping the boards of the great hall with a short broom, and in an attitude closely approaching all-fours, when Saï, who was hidden under one of the sofas, suddenly leaped upon her back, where he stood waving his tail in triumph. She screamed so violently as to summon the other servants, but they, seeing the panther in the act of devouring her, as they thought, gallantly scampered off one and all, as fast as their heels could carry them; nor was the woman released from her load till the governor, hearing the noise, came to her assistance.

Mrs. Bowdich determined to take this interesting animal to England, and he was conveyed on board ship, in a large wooden cage thickly barred in front with iron. Even this confinement was not deemed a sufficient protection by the canoe men, who were so alarmed that in their confusion they managed to drop cage and all into the sea. For a few minutes the poor fellow was given up for lost, but some sailors jumped into a boat belonging to the vessel, and dragged him out in safety. He seemed completely subdued by his ducking; and as no one dared to open the cage to dry it, he rolled himself up in one corner, where he remained for some days, till roused by the voice of his mistress. When she first spoke, he raised his head, listened attentively, and when she came fully into view, he jumped on his legs, and appeared frantic, rolling over and over, howling and seeming as if he would have torn his cage to pieces; however, his violence gradually subsided, and he contented himself with thrusting his nose and paws through the bars to receive her caresses. The greatest treat that could be bestowed on Saï was lavender-water. Mr. Hutchinson had told Mrs. Bowdich, that on the way from Ashantee, happening to draw out a scented pocket-handkerchief, it was immediately seized by the panther, who reduced it to atoms; nor could he venture to open a bottle of perfume when the animal was near, he was so eager to enjoy it. Twice a week his mistress indulged him by making a cup of stiff paper, pouring a little lavender-water into it, and giving it to him through the bars of the cage; he would drag it to him with great eagerness, roll himself over it, nor rest till the smell had evaporated.

Quiet and gentle as Saï was, pigs never failed to excite indignation when they hovered about his cage, and the sight of a monkey put him in a complete fury. While at anchor in the 'Gaboon,' an orang-outang was brought on board, and remained three days. When the two animals met, the uncontrollable rage of the one and the agony of the other was very remarkable. The orang was about three feet high, and very powerful, so that when he fled with extraordinary rapidity from the panther to the other side of the deck, neither men nor things remained upright if they opposed his progress. As for the panther, his back rose in an arch, his tail was elevated and perfectly stiff, his eyes flashed, and as he howled he showed his huge teeth; then, as if forgetting the bars before him, he made a spring at the orang to tear him to atoms. It was long before he recovered his tranquillity; day and night he was on the listen, and the approach of a monkey or a negro brought back his agitation. During the voyage to England the vessel was boarded by pirates, and the crew and the passengers nearly reduced to starvation in consequence; Saï must have died had it not been for a collection of more than three hundred parrots; of these his allowance was one per diem, but he became so ravenous that he had not patience to pick off the feathers, but bolted the birds whole; this made him very ill, but Mrs. Bowdich administered some pills, and he recovered. On the arrival of the vessel in the London Docks, Saï was presented to the Duchess of York, who temporarily placed him in Exeter 'Change. On the morning of the Duchess's departure for Oatlands, she went to visit her new pet, played with him, and admired his gentleness and great beauty. In the evening, when her royal highness's coachman went to take him away to his new quarters at Oatlands, Saï was dead from inflammation on the lungs.

To this interesting animal the following lines, by Dryden, might with propriety have been applied:—

"The Panther, sure the noblest next the Hind
And fairest creature of the spotted kind;
Oh, could her inborn stains be washed away,
She were too good to be a beast of prey!
How can I praise or blame, and not offend,
Or how divide the frailty from the friend?
Her faults and virtues lie so mix'd that she
Nor wholly stands condemned, nor wholly free."

Mr. Gordon Cumming describes two encounters with leopards, one of which was nearly attended with fatal consequences:—"On the 17th," says he, "I was attacked with acute rheumatic fever, which kept me to my bed, and gave me excruciating pain. Whilst I lay in this helpless state, Mr. Orpen and Present, who had gone up the river to shoot sea-cows, fell in with an immense male leopard, which the latter wounded very badly. They then sent natives to camp to ask me for dogs, of which I sent them a pair. In about an hour the natives came running to camp, and said that Orpen was killed by the leopard. On further inquiry, however, I found that he was not really killed, but frightfully torn and bitten about the arms and head. They had rashly taken up the spoor on foot, the dogs following behind them, instead of going in advance. The consequence of this was, that they came right upon the leopard before they were aware of him, when Orpen fired, and missed him. The leopard then sprang on his shoulders, and dashing him to the ground lay upon him, howling and lacerating his hands, arms and head most fearfully. Presently the leopard permitted Orpen to rise and come away. Where were the gallant Present and all the natives, that not a man of them moved to assist the unfortunate Orpen? According to an established custom among all colonial servants, the instant the leopard sprang, Present discharged his piece in the air, and then dashing it to the ground, he rushed down the bank and jumped into the river, along which he swam some hundred yards before he would venture on *terra firma*. The natives, though numerous and armed, had likewise fled in another direction."

The tenacity of life of these animals was well shown in the other encounter:—" Having partaken of some refreshment," says Mr. Cumming, "I saddled two steeds, and rode down the banks of Ngotwani, with the Bushman, to seek for any game I might find. After riding about a mile along the river's bank, I came suddenly upon an old male leopard lying under the shade of a thorn grove, and panting from the great heat. Although I was within sixty yards of him he had not heard the horse's tread. I thought he was a lioness, and dismounting, took a rest in my saddle on the old grey, and sent a bullet into him. He sprang to his feet, and ran halfway down the river's bank, and stood to look about him, when I sent a second bullet into his person, and he disappeared over the bank. The ground being very dangerous, I did not disturb him by following then, but I at once sent Ruyter back to camp for the dogs.

"Presently he returned with Wolf and Boxer, very much done up with the sun. I rode forward, and on looking over the bank, the leopard started up and sneaked off alongside of the tall reeds, and was instantly out of sight. I fired a random shot from the saddle, to encourage the dogs, and shouted to them; they, however, stood looking stupidly round, and would not take up his scent at all. I led them over his spoor again and again, but to no purpose; the dogs seemed quite stupid, and yet they were Wolf and Boxer, my two best. At length I gave it up as a lost affair, and was riding down the river's bank, when I heard Wolf give tongue behind me, and galloping back I found him at bay, with the leopard immediately beneath where I had first fired at him; he was very severely wounded, and had slipped down into the river's bed, and doubled back, whereby he had thrown out both dogs and myself. As I approached, he flew out upon Wolf and knocked him over, and then running up the bed of the river, he took shelter in a thick bush. Wolf, however, followed him, and at this moment my other dogs came up, having heard the shot, and bayed him fiercely. He sprang out upon them, and then crossed the river's bed, taking shelter beneath some large tangled roots on the opposite bank. As he crossed the river, I put a third bullet into him, firing from the saddle, and as soon as he came to bay I gave him a fourth, which finished him. This leopard was a very fine old male. In the conflict the unfortunate Alert was wounded as usual, getting his face torn open. He was still going on three legs, with all his breast laid bare by the first waterbuck."

Major Denham gives the following account of an adventure with a huge panther, which occurred during the expedition to Mandara:—" We had started several animals of the leopard species, who ran from us so swiftly, twisting their long tails in the air, as to prevent our getting near them. We however now started one of a larger kind, which Maramy assured me was so satiated with the blood of a negro, whose carcase we found lying in the wood, that he would be easily killed. I rode up to the spot just as Shonaa had planted the first spear in him, which passed through the neck a little above the shoulder, and came down between the animal's legs; he rolled over, broke the spear, and bounded off with the lower half in his body. Another Shonaa galloped up within two arms' length and thrust a second through his loins; and the savage animal, with a woful howl, was in the act of springing on his person, when an Arab shot him through the head with a ball which killed him on the spot. It was a male panther of a very large size, and measured, from the point of the tail to the nose, eight feet two inches."

These animals are found in great abundance in the woods bordering on Mandara; there are also leopards, the skins of which were seen, but not in great numbers. The panthers are as insidious as they are cruel; they will not attack anything that is likely to make resistance, but have been known to watch a child for hours while near the protection of huts or people. They will often spring on a grown person, male or female, while carrying a burden, but always from behind. The flesh of a child or young kid they will sometimes devour, but when any full-grown animal falls a prey to their ferocity, they suck the blood alone.

We are indebted to a very distinguished officer, Major W. T. Johnson, late commanding the 12th Irregular Cavalry, for interesting particulars relative to panthers and bears in India.

"Bears generally go in couples. On one occasion I became aware of some rocks where bears were known to live, but there was no getting them out. Knowing that they always go out to feed at night and return about sunrise, I went before dawn to the rocks, and had hardly taken up my position, before I saw them coming one after the other. I took a shot at the leading one, and hit him somewhere about the tail; he was probably under the impression that his companion had bitten him in the stern, for he turned round and furiously seized him; they both rolled over and had a most terrible battle: it was thoroughly ludicrous.

"I have seldom known tigers or bears to charge without provocation, or being wounded. I have seen and shot many bears, and on only one occasion have I known them commence the assault: it was when a party of men were going through some rocks; a bear came out suddenly, seized one of the beaters, shook him well, put him down, and walked off: I was not on the spot, and the bear got away. Bears sometimes carry off a wonderful deal of lead; I have shot them through and through, yet they have made their way into the jungle and escaped.

"I believe there is not a more treacherous or dangerous animal than a panther: many more accidents happen with panthers than with tigers in India. On one occasion we had wounded the mother of some cubs, and we went after her with great caution, well armed with guns, spears, and shields, all covering the *puggee*, or tracker, who went close in front of us; she was waiting, and charged straight on us from behind the root of a tree; I jumped on one side, gave her a shot as she went past, and hit her in the neck. She charged on from one man to another, each having a shot, or a cut at her with his sword; I never saw a more game brute in my life, and she continued to charge from one to the other till she had not breath left in her body.

"We took the cubs home, and I brought one up, and a voracious little savage he proved; he used to hunt the goats and poultry, and finished up by devouring part of a pair of flannel *pants* and a leather razor-case. I had observed the brute getting poorly, and administered two grains of tartar emetic and two of calomel, which caused him to vomit up the above articles.

"There was once marked down for me, in a small patch of bushes, a large panther which I knew to be very severely wounded, and I thought disabled. I took my rifle, a handy double-barrelled Lancaster, and walked towards the bushes; when I was within forty yards, he came out at me so quickly that I had but just time to put up my rifle and fire both barrels as quick as I could. Most fortunately one bullet entered just above the left eye, and came out behind the ear somewhat confusing him, but not in the least checking his speed; he knocked me over and bit and clawed me severely. Some of my men, of the Goozerat Irregular Horse, behaved very well, and attacking him drove him off me with their swords and carbines. We killed him at last, but I had a very narrow escape."

Lichtenstein* describes an interesting scene of

* Travels in Southern Africa, by H. Lichtenstein, 1815.

which he was an eye-witness, by which we learn the manner in which leopards are taken alive in Africa. "One of the colonists having, just at this time, caught a large leopard, sent round to his friends to inform them of it, inviting them, according to the custom of the country, to assemble on a day appointed, in the afternoon, to see the combat between this animal and dogs. After partaking of an excellent dinner, we were conducted to the snare where the creature was still confined, whence he must be taken very cautiously to be carried to the place of combat. This snare was in the remote part of a mountain dell, and was enclosed by a wall of rough pieces of stone, so that two large blocks like the others formed the entrance. For the rest, with regard to the mechanism, it was constructed upon the same principle as a mousetrap, only with the proper difference of proportions. The snares made for hyenas are of a similar construction, excepting that they are open above; this, on the contrary, was covered with rough planks, between which we could look down on the beautiful and enraged beast, and on which stood the people who were now to fetter him. They began with throwing in slings, by which first one paw, then another, was caught, and the legs were thus drawn together, while he in vain raged and roared most terribly. When this was done, another person went in who threw a sling over the head, by the assistance of which the creature was half drawn out. A strong muzzle was then tied over his mouth, and, thus secured, he was carried to the place of combat. A cord was now thrown round his body just above the haunches, to which a chain was affixed, and that was fastened to a strong post. By degrees his bandages were taken off, and at length he was left with no other confinement than being tethered to the post. He soon recovered his strength and agility, and began alternately his wild springs, and his graceful movements to and fro, exhibiting, indeed, a very fine spectacle: it was one of which no person can have an idea who has only seen these creatures in cages where they are shown about by the exhibitors of wild beasts in our own country, humbled and tamed as they are by chastisement, hunger, and the damp cold of a European climate.

"This South African leopard differs from that of Northern Africa,—the true panther,—in the form of its spots, in the more slender structure of its body, and in the legs not being so long in proportion to the body. In watching for his prey he crouches on the ground with his fore paws stretched out, and his head between them, his eyes rather directed upwards. In this manner he now laid himself down, and being held fast by the chain, stretched himself to such a length that he appeared entirely a different animal. He then unexpectedly twined his body about sideways, so that his movements very much resembled those of a snake. Convinced that he was sufficiently secured by the chain, we ventured close to him—we even sought to tease and provoke him to spring and roar by throwing little stones at him, and by playing other tricks. As evening was, however, coming on, a consultation was held whether it would not be advisable to set the dogs upon him, for hitherto they had been kept in confinement, that we might at first see as much as we wished of the manners and behaviour of the prisoner. The question being determined in the affirmative, most of the company went to prepare these new combatants for the field; when the leopard, making a grand effort, broke his chain, and being thus left entirely at liberty made a formidable spring at the landrost and me, who had ventured rather too near him. We took to flight under the utmost alarm and astonishment; but happily the leopard's strength being somewhat exhausted, he missed his aim, and at that important moment, before he could attempt a second spring, the dogs, who were now let loose, rushed upon him, and immediately seized him by the throat and ears. One of them, who had from age lost a tooth upon our journey, was easily shaken off by the monster, who killed him instantly by a desperate bite on the head. The rest of the dogs now fell furiously upon him, and two of them bit him in the throat so effectually, that in less than a quarter of an hour not the least spark of remaining life was to be discerned. On dissecting the animal I found all the muscles about the throat and neck bit, but not the slightest hole made in the skin. As it was wholly uninjured, I purchased it of the farmer at the usual price given here for leopards' skins—ten dollars."

In India and Ceylon leopards and panthers are called Tree Tigers, and the following narrative of an exciting encounter with one is given in 'The Menageries.' "I was at Jaffna," says the writer, 'at the northern extremity of the island of Ceylon, in the beginning of the year 1819, when one morning my servant called me an hour or two before the usual time with, 'Master! master! people sent for master's dogs; tiger in the town!' Now my dogs chanced to be very degenerate specimens of a fine species called the Poligar dogs. I kept them to hunt jackals, but tigers are very different things. This turned out to be a panther. My gun chanced not to be put together; and while my servant was doing it, the collector and two medical men, who had recently arrived, came to my door, the former armed with a fowling-piece, and the two latter with remarkably blunt hogspears. They insisted on setting off without waiting for my gun, a proceeding not much to my taste. The tiger

(I must continue to call him so) had taken refuge in a hut, the roof of which, as those of Ceylon huts in general, spread to the ground like an umbrella; the only aperture was a small door about four feet high. The collector wanted to get the tiger out at once. I begged to wait for my gun, but, no! the fowling-piece, loaded with ball of course, and the two hogspears were quite enough. I got a hedge-stake and awaited my fate for very shame. At that moment, to my great delight, there arrived from the fort an English officer, two artillerymen, and a Malay captain, and a pretty figure we should have cut without them, as the event will show. I was now quite ready to attack, and my gun came a minute afterwards. The whole scene which follows took place within an enclosure, about twenty feet square, formed on three sides by a strong fence of palmyra leaves, and on the fourth by the hut. At the door of this the two artillerymen planted themselves, and the Malay captain got at the top to frighten the tiger out by worrying it—an easy operation, as the huts there are covered with cocoa-nut leaves. One of the artillerymen wanted to go in to the tiger, but we would not suffer it. At last the beast sprang; this man received him on his bayonet, which he thrust, apparently, down his throat, firing his piece at the same moment. The bayonet broke off short, leaving less than three inches on the musket;

the rest remained in the animal, but was invisible to us: the shot probably went through his cheek, for it certainly did not seriously injure him, as he instantly rose upon his legs with a loud roar, and placed his paws upon the soldier's breast. At this moment the animal appeared to me to be about to reach the centre of the man's face; but I had scarcely time to observe this, when the tiger, stooping his head, seized the soldier's arm in his mouth, turned him half round staggering, threw him over on his back, and fell upon him. Our dread now was, that if we fired upon the tiger, we might kill the man. For a moment there was a pause, when his comrade attacked the beast exactly in the same manner the gallant fellow himself had done. He struck his bayonet into his head; the tiger rose at him, he fired, and this time the ball took effect, and in the head. The animal staggered backwards, and we all poured in our fire; he still kicked and writhed, when the gentlemen with the hogspears advanced and fixed him, while some natives finished him by beating him on the head with hedge-stakes. The brave artilleryman was after all but slightly hurt: he claimed the skin, which was very cheerfully given to him; there was, however, a cry among the natives, that the head should be off: it was, and in doing so, the knife came directly across the bayonet. The animal measured scarcely less than four feet from the root of the tail to the muzzle."

The following practical joke is related in the late Rev. T. Acland's amusing volume on India:—" A party of officers went out from Cuttack to shoot: their men were beating the jungle, when suddenly all the wild cry ceased, and a man came gliding to where all the Sahibs were standing, to tell them that there was a tiger lying asleep in his den close at hand. A consultation was instantly held; most of the party were anxious to return to Cuttack, but Captain B—— insisted on having a shot at the animal; accordingly he advanced very quickly, until he came to the place, when he saw, not a tiger, but a large leopard, lying quite still, with his head resting on his fore paws. He went close up and fired, but the animal did not move. This astonished him, and on examination he found that the brute was already dead. One of his companions had bribed some Indians to place a dead leopard there, and to say that there was a tiger asleep. It may be imagined what a laugh there was!"

Nature, ever provident, has scattered with a bounteous hand her gifts in the country of the Orinoco, where the jaguar especially abounds. The Savannahs, which are covered with grasses and slender plants, present a surprising luxuriance and diversity of vegetation; piles of granite blocks rise here and there, and at the margins of the plains occur deep valleys and ravines, the humid soil of which is covered with arums, heliconias, and llianas. The shells of primitive rocks, scarcely elevated above the plain, are partially coated with lichens and mosses, together with succulent plants, and tufts of evergreen shrubs with shining leaves. The horizon is bounded with mountains overgrown with forests of laurels, among which clusters of palms rise to the height of more than a hundred feet, their slender stems supporting tufts of feathery foliage. To the east of Atures other mountains appear, the ridge of which is composed of pointed cliffs, rising like huge pillars above the trees. When these columnar masses are situated near the Orinoco, flamingoes, herons, and other wading birds perch on their summits, and look like sentinels. In the vicinity of the cataracts, the moisture which is diffused in the air produces a perpetual verdure, and wherever soil has accumulated on the plains, it is adorned by the beautiful shrubs of the mountains.

Such is one view of the picture, but it has its dark side also; those flowing waters, which fertilize the soil, abound with crocodiles; those charming shrubs and flourishing plants, are the hiding-places of deadly serpents; those laurel forests, the favourite lurking-spots of the fierce jaguar; whilst the atmosphere, so clear and lovely, abounds with mosquitoes and zancudoes, to such a degree that, in the missions of Orinoco, the first questions in the morning when two people meet, are "How did you find the zancudoes during the night? How are we to-day for the mosquitoes?"

It is in the solitude of this wilderness that the jaguar, stretched out, motionless and silent, upon one of the lower branches of the ancient trees, watches for its passing prey; a deer, urged by thirst, is making its way to the river, and approaches the tree where his enemy lies in wait. The jaguar's eyes dilate, the ears are thrown down, and the whole frame becomes flattened against the branch. The deer, all unconscious of danger, draws near,—every limb of the jaguar quivers with excitement; every fibre is stiffened for the spring; then, with the force of a bow unbent, he darts with a terrific yell upon his prey, seizes it by the back of the neck, a blow is given with his powerful paw, and with broken spine the deer falls lifeless to the earth. The blood is then sucked, and the prey dragged to some favourite haunt, where it is devoured at leisure.

Humboldt surprised a jaguar in his retreat. It was near the Joval, below the mouth of the Cano de la Tigrera, that in the midst of wild and awful scenery, he saw an enormous jaguar stretched beneath the shade of a large mimosa. He had just killed a chiguire, an animal about the size of a pig, which he held with one of his paws, while the vultures were assembled in flocks around. It was curious to observe the mixture of boldness and timidity which these birds exhibited; for although they advanced within two feet of the jaguar, they instantly shrank back at the least motion he made. In order to observe more nearly their proceedings, the travellers went into their little boat, when the tyrant of the forest withdrew behind the bushes, leaving his victim, upon which the vultures attempted to devour it, but were soon put to flight by the jaguar rushing into the midst of them; the following night Humboldt and his party were entertained by a jaguar-hunter, half-naked, and as brown as a Zambo, who prided himself on being of the European race, and called his wife and daughter, who were as slightly clothed as himself, Donna Isabella and Donna Manvela. As this aspiring personage had neither house nor hut, he invited the strangers to swing their hammocks near his own between two trees; but as ill-luck would have it, a thunder-storm came on, which wetted them to their skin; but their troubles did not end here, for Donna Isabella's cat had perched on one of the trees, and frightened by the thunder-storm, jumped down upon one of the travellers in his cot; he naturally supposed that he was attacked by a wild beast, and as smart a battle took place between the two, as that celebrated feline engagement of Don Quixote; the cat, who perhaps had most reason to consider himself an ill-used personage, at length bolted, but the fears of the gentleman had been excited to such a degree, that he could hardly be quieted. The following night was not more propitious to slumber. The party finding no tree convenient, had stuck their oars in the sand, and suspended their hammocks upon them. About eleven, there arose in the immediately adjoining wood, so terrific a noise, that it was impossible to sleep. The Indians distinguished the cries of sapagous, alouates, jaguars, cougars, peccaris, sloths, curassows, paraquas, and other birds, so that there must have been as full a forest chorus as Mr. Hullah himself could desire.

When the jaguars approached the edge of the forest, which they frequently did, a dog belonging to the party began to howl, and seek refuge under their cots. Sometimes, after a long silence, the cry of the jaguars came from the tops of the trees, when it was followed by an outcry among the monkeys. Humboldt supposes the noise thus made by the inhabitants of the forest during the night, to be the effect of some contests that have arisen among them.

On the pampas of Paraguay, great havoc is committed among the herds of horses by the jaguars, whose strength is quite sufficient to enable them to drag off one of these animals. Azara caused the body of a horse, which had been recently killed by a jaguar, to be drawn within musket-shot of a tree, in which he intended to pass the night, anticipating that the jaguar would return in the course of it, to its victim; but while he was gone to prepare for his adventure, behold the animal swam across a large and deep river, and having seized the horse with his teeth, dragged it full sixty paces to the river, swam across again with his prey, and then dragged the carcase into a neighbouring wood; and all this in sight of a person whom Azara had placed to keep watch. But the jaguars have also an aldermanic *goût* for turtles, which they gratify in a very systematic manner, as related by Humboldt, who was shown large shells of turtles emptied by them. They follow the turtles towards the beaches, where the laying of eggs is to take place, surprise them on the sand, and in order to devour them at their ease, adroitly turn them on their backs; and as they turn many more than they can devour in one night, the Indians often profit by their cunning. The jaguars pursue the turtle quite into the water, and when not very deep, dig up the eggs: they, with the crocodile, the heron, and the gallinago vulture, are the most formidable enemies the little turtles have. Humboldt justly remarks, "When we reflect on the difficulty that the naturalist finds in getting out the body of the turtle, without separating the upper and under shells, we cannot enough admire the suppleness of the jaguar's paw, which empties the double armour of the *arraus*, as if the adhering parts of the muscles had been cut by means of a surgical instrument."

The rivers of South America swarm with crocodiles, and these wage perpetual war with the jaguars. It is said, that when the jaguar surprises the alligator asleep on the hot sand-bank, he attacks him in a vulnerable part under the tail, and often kills him; but let the crocodile only get his antagonist into the water, and the tables are turned, for the jaguar is held under water until he is drowned.

The onset of the jaguar is always made from behind, partaking of the stealthy, treacherous character of his tribe; if a herd of animals or a party of men be passing, it is the last that is always the object of his attack. When he has made choice of his victim, he springs upon the neck, and placing one paw on the back of the head, while he seizes the muzzle with the other, twists the head round with a sudden jerk which dislocates the spine, and deprives it instantaneously of life; sometimes, especially when satiated with food, he is indolent and cowardly, skulking in the gloomiest depths of the forest, and scared by the most trifling causes, but when urged by the cravings of hunger, the largest quadrupeds, and man himself, are attacked with fury and success.

Mr. Darwin* has given an interesting account of the habits of the jaguars: the wooded banks of the great South American rivers appear to be their fa-

---

* Researches in Geology and Natural History.

vourite haunt, but south of the Plata they frequent the reeds bordering lakes; wherever they are, they seem to require water. They are particularly abundant on the isles of the Parana, their common prey being the carpincho, so that it is generally said, where carpinchos are plentiful, there is little fear of the jaguar; possibly, however, a jaguar which has tasted human flesh, may afterwards become dainty, and, like the lions of South Africa, and the tigers of India, acquire the dreadful character of a man-eater, from preferring that food to all others. It is not many years ago since a very large jaguar found his way into a church in Santa Fé; soon afterwards a very corpulent padre entering, was at once killed by him: his equally stout coadjutor, wondering what had detained the padre, went to look after him, and also fell a victim to the jaguar; a third priest, marvelling greatly at the unaccountable absence of the others, sought them, and the jaguar, having by this time acquired a strong clerical taste, made at him also, but he, being fortunately of the slender order, dodged the animal from pillar to post, and happily made his escape. The beast was destroyed by being shot from a corner of the building, which was unroofed, and thus paid the penalty of his sacrilegious propensities.

On the Parana they have killed many woodcutters, and have even entered vessels by night. One dark evening the mate of a vessel, hearing a heavy but peculiar footstep on deck, went up to see what it was, and was immediately met by a jaguar, who had come on board, seeking what he could devour: a severe struggle ensued, assistance arrived, and the brute was killed, but the man lost the use of an arm which had been ground between his teeth.

The Guachos say that the jaguar, when wandering about at night, is much tormented by the foxes yelping as they follow him. This may perhaps serve to alarm his prey, but must be as teasing to him as the attentions of swallows are to an owl who happens to be taking a daylight promenade; and if owls ever swear, it is under those circumstances. Mr. Darwin, when hunting on the banks of the Uruguay, was shown three well-known trees to which the jaguars constantly resort, for the purpose, it is said, of sharpening their claws. Every one must be familiar with the manner in which cats with outstretched legs and extended claws, will card the legs of chairs and of men; so with the jaguar; and of these trees, the bark was worn quite smooth in front; on each side there were deep grooves, extending in an oblique line nearly a yard in length. The scars were of different ages, and the inhabitants could always tell when a jaguar was in the neighbourhood, by his recent autograph on one of these trees.

We have seen tigers stretching their enormous limbs

in this manner, and were interested in watching the proceedings of two beautiful young jaguars in the Zoological Gardens, Regent's Park; they were scarcely half grown, and as playful as kittens. After chasing and tumbling each other over several times, they went as by mutual consent to the post of their cage, and there carefully, and with intensely placid countenances, scraped away with their claws as they would have done against the trees had they been in their native woods. This proceeding satisfactorily concluded, they swarmed up and down the post, appearing to vie with each other as to which should be first. The young leopards were equally graceful and active with the jaguars, and the elegance and quickness of their movements never failed to command admiration. They seemed to be particularly fond of bounding up and down the trees, and sometimes rested in the strangest attitudes, stuck in the fork of a bough, or sitting as it were astride of one, with their hind legs hanging down. M. Sonnini bears testimony to the extraordinary climbing powers of the jaguar: "For," says he, "I have seen, in the forests of Guiana, the prints left by the claws of the jaguar on the smooth bark of a tree from forty to fifty feet in height, measuring about a foot and a half in circumference, and clothed with branches near its summit alone. It was easy to follow with the eye the efforts which the animal had made to reach the branches; although his talons had been thrust deeply into the body of the tree, he had met with several slips, but had always recovered his ground; and attracted, no doubt, by some favourite object of prey, had at length succeeded in gaining the very top!"

The following is the common mode of killing the jaguar in Tucuman:—The Guacho, armed with a long strong spear, traces him to his den, and having found it, places himself in a convenient position to receive the animal on the point of his spear at the first spring; dogs are then sent in, and driving him out, he springs with fury upon the Guacho, who, fixing his eyes on those of the jaguar, receives his onset kneeling, and with such consummate coolness that he scarcely ever fails. At the moment that the spear is plunged into the animal's body the Guacho nimbly springs on one side, and the jaguar, already impaled on the spear, is speedily despatched.

In one instance the animal lay stretched on the ground, like a gorged cat, and was in such high good-humour after his satisfactory meal, that on the dogs attacking him he was disposed to play with them; a bullet was therefore lodged in his shoulder, on which rough salute he sprang out so quickly on his watching assailant, that he not only received the spear in his body but tumbled the man over, and they rolled on the ground together. "I thought," said the brave fellow, "that I was no longer a capitaz, as I held up my arm to protect my throat, which the jaguar seemed in the act of seizing; but at the very moment that I expected to feel his fangs in my flesh, the green fire which had blazed upon me from his eyes flashed out— he fell upon me, and with a quiver died."

Colonel Hamilton* relates that, when travelling on the banks of the Magdalena, he remarked a young man with his arm in a sling, and on inquiring the cause, was told, that about a month before, when walking in a forest, a dog he had with him began to bark at something in a dark cavern overhung with bushes; and on his approaching the entrance, a jaguar rushed on him with great force, seizing his right arm, and in the struggle they both fell over a small precipice. He then lost his senses, and, on recovering, found the jaguar had left him, but his arm was bleeding, and shockingly lacerated. On surprise being expressed that the animal had not killed him, he shrugged up his shoulders, and remarked, "La bienaventurada virgen Maria le habia salvo,"—"The blessed Virgin had saved him."

In the province of Buenaventura it is said that the Indians kill the jaguar by means of poisoned arrows, about eight inches in length, which are projected from

* Travels through the Interior Provinces of Columbia. London, 1827.

a blow-pipe: the arrows are poisoned with a moisture which exudes from the back of a small green frog, found in the provinces of Buenaventura and Choco. When the Indians want to get this poison from the frog, they put him near a small fire, and the moisture soon appears on his back; in this the points of the small arrows are dipped, and so subtle is the poison, that a jaguar struck by one of these little insignificant weapons soon becomes convulsed, and dies.

The jaguar has the general character of being untameable, and of maintaining his savage ferocity when in captivity, showing no symptoms of attachment to those who have the care of them. This, like many other points in natural history, is a popular error: there is at the present time (1852) a magnificent jaguar in the Zoological Gardens, who is as tame and gentle as a domestic cat. We have seen this fine creature walking up and down the front of his den as his keeper walked, rubbing himself against the bars, purring with manifest pleasure as his back or head was stroked, and caressing the man's hand with his huge velvet paws. There is in the collection another jaguar, just as savage as this one is tame. There was a jaguar formerly in the Tower, which was obtained by Lord Exmouth while on the South American station, and was afterwards present at the memorable bombardment of Algiers: this animal was equally gentle with that we have described, and was presented to the Marchioness of Londonderry by Lord Exmouth on his return to England after that engagement: it was placed by her Ladyship in the Tower, where it died.

In a state of nature these animals have been known to show not only forbearance, but even playfulness, of which Humboldt relates the following instance which occurred at the mission of Atures, on the banks of the Orinoco:—" Two Indian children, a boy and a girl, eight or nine years of age, were sitting among the grass near the village of Atures, in the midst of a savannah. It was two in the afternoon when a jaguar issued from the forest and approached the children, gambolling round them, sometimes concealing himself among the long grass, and again springing forward with his back curved and his head lowered, as is usual with our cats. The little boy was unaware of the danger in which he was placed, and became sensible of it only when the jaguar struck him on the side of the head with one of his paws. The blows thus inflicted were at first slight, but gradually became ruder; the claws of the jaguar wounded the child, and blood flowed with violence; the little girl then took up the branch of a tree and struck the animal, which fled before her. The Indians, hearing the cries of the children, ran up, and saw the jaguar, which bounded

off without showing any disposition to defend itself."*
In all probability this fit of good humour was to be
traced to the animal having been plentifully fed; for
most assuredly the children would have stood but little
chance had their visitor been subjected to a meagre
diet for some days previously.

Mr. Edwards, in his Voyage up the Amazon, tells of
an exchange of courtesies between a traveller and a
jaguar. The jaguar was standing in the road as the
Indian came out of the bushes, not ten paces distant,
and was looking, doubtless, somewhat fiercely as he
waited the unknown comer. The Indian was puzzled
for an instant, but summoning his presence of mind,
he took off his broad-brimmed hat, and made a low
bow, with "Muito bene dias, meu Senhor," or, "A
very good morning, Sir." Such profound respect was
not wasted on the jaguar, who turned slowly and
marched down the road with proper dignity.

It is difficult to say how many leopard and jaguar
skins are annually imported, as the majority are
brought by private hands. We have been told by an
eminent furrier that about five hundred are sold each
year to the London trade. They are chiefly used as
shabraques, or coverings to officers' saddles in certain
hussar regiments; but skins used for this purpose
must be marked in a particular manner, and the ground

* Travels and Researches of Alexander Von Humboldt, 1851.

must be of a dark rich colour. Such skins are worth about three pounds; ordinary leopard and jaguar skins are valued at about two pounds, and are chiefly used for rugs or mats. The jaguar-skins are sometimes of great size, and we have measured one which was nine feet seven inches from tip to tip. The leopard-skins are exclusively used for military purposes, and the jaguars' are preferred for rugs.

CHAPTER IV.

WOLVES.—ORIGIN OF DOGS AND WOLVES.—POINTS OF DIFFERENCE.—VARIETIES.—ANCIENT SUPERSTITIONS.—MYTHS.—CHARMS.—WOLVES IN BRITAIN.—ANCIENT LAWS.—THE LAST WOLF.—CUNNING.—AN UNWELCOME VISITOR.—A CLEVER PERFORMER.—A SOCIABLE ANIMAL.—VALUE OF SKINS.—CROSS BREED.—AN ADVENTURE.—TRAPS.—WOLF AND REINDEER.—BOLDNESS.—DOGS KILLED BY WOLVES.—CUNNING OF FOXES.—A MIDNIGHT STRUGGLE.—AN AFFECTIONATE WOLF.—TUSSA.—THEFT OF AN INFANT.—TASTE FOR PORK.—AN AWKWARD PREDICAMENT.—A FIERCE PIG.—A SOLDIER DEVOURED.—CRY OF THE JACKAL.—THE OLD QUARTERMASTER.—AN ATTRACTIVE FIGURE.—THE SAILOR AND THE BEEF.—UTILITY OF WOLVES.

A PECULIAR interest attaches to the wolf, from the close analogy which in all its essential features it presents to the faithful companion of man. So close, indeed, is the analogy, that some of the ablest zoologists, the celebrated John Hunter included, have entertained the opinion that dogs, in all their varieties, and wolves, have descended from a common stock. With the exception of an obliquity in the position of the eyes, there is no appreciable anatomical difference between these animals. The question is one of difficulty; but we believe we are correct in stating that

the majority of the highest authorities agree in the
belief that these animals are not derived from a common
parent, but were originally distinct, and will ever
so continue. There are several species of wild dogs
known, quite distinct from the wolf; and although the
opportunities have been numerous for dogs resum‑
ing their pristine form by long continuance in a sa‑
vage state, no instance has ever occurred of their be‑
coming wolves, however much they might degenerate
from the domestic breed. The honest and intelligent
shepherd-dog was regarded by Buffon as the "*fons et
origo*" from which all other dogs, great and small,
have sprung, and he drew up a kind of genealogical
table, showing how climate, food, education, and inter‑
mixture of breeds gave rise to the varieties. At Kat‑
mandoo there are many plants found in a wild state,
which man has carried with him in his migrations, and
wild animals, which may present the typical forms
whence some of our domestic races have been derived;
among these is a wild dog, which Mr. Hodgson con‑
siders to be the primitive species of the whole canine
race. By Professor Kretchner, the jackal was re‑
garded as the type of the dogs of ancient Egypt, an
idea supported by the representations on the walls
of the temples. This question, however, of the origin
of the canine race, is so thoroughly obscured by the
mists of countless ages, as to be incapable of direct

proof. Philosophers may indulge themselves with speculations; but in the absence of that keystone, proof, the matter must rest on the basis of theory alone.

The following are some of the chief differences between wolves, wild dogs, and domestic dogs. The ears of the wild animals are always pricked, the lop or drooping ear being essentially a mark of civilization; with very rare exceptions, their tails hang more or less and are bushy, the honest cock of the tail so characteristic of a respectable dog, being wanting. This is certainly the rule; but, curious enough, the Zoological Gardens contain at the present moment, a Portuguese female wolf which carries her tail as erect and with as bold an air as any dog. Wolves and wild dogs growl, howl, yelp, and cry most discordantly, but, with one exception, do not bark; that exception being the wild hunting-dog of South Africa, which, according to Mr. Cumming, has three distinct cries; one is peculiarly soft and melodious, but distinguishable at a great distance: this is analogous to the trumpet-call, "halt and rally," of cavalry, serving to collect the scattered pack when broken in hot chase. A second cry, which has been compared to the chattering of monkeys, is emitted at night when the dogs are excited; and the third note is described as a sharp, angry bark, usually uttered when they behold an object they cannot make out, but which differs from the true, well-known bark of the domestic dog.

The common or European wolf is found from Egypt to Lapland, and is most probably the variety that formerly haunted these islands. The wolves of Russia are large and fierce, and have a peculiarly savage aspect. The Swedish and Norwegian are similar to the Russian in form, but are lighter in colour, and in winter totally white. Those of France are browner and smaller than either of these, and the Alpine wolves are smaller still. Wolves are very numerous in the northern regions of America; "their foot-marks," says Sir John Richardson, "may be seen by the side of every stream, and a traveller can rarely pass the night in these wilds without hearing them howling around him."* These wolves burrow, and bring forth their young in earths with several outlets, like those of the fox. Sir John saw none with the gaunt appearance, the long jaw and tapering nose, long legs and slender feet, of the Pyrenean wolves.

India, too, is infested with wolves, which are smaller than the European. There is a remarkably fine animal at the Zoological Gardens, born of a European father and Indian mother, which, in size and other respects, so closely partakes of the characteristics of his sire, that he might well pass for pure blood.

Among the ancients, wolves gave rise to many superstitious fictions. For instance, it was said that they

* Fauna Boreali-Americana, p. 62.

possessed "an evil eye," and that, if they looked on a man before he saw them, he would forthwith lose his voice. Again, we find the Roman witches, like the weird sisters of Macbeth, employing the wolf in their incantations:—

> "Utque lupi barbam variæ cum dente colubræ
> Abdiderint furtim terris."—HOR., *Sat.* viii. lib. i.

There was a myth prevalent among the ancients, that in Arcadia there lived a certain family of the Antæi, of which one was ever obliged to be transformed into a wolf. The members of the family cast lots, and all accompained the luckless wight on whom the lot fell, to a pool of water. This he swam over, and having entered into the wilderness on the other side, was forthwith in form a wolf, and for nine years kept company with wolves; at the expiration of that period he again swam across the pool, and was restored to his natural shape, only that the addition of nine years was placed upon his features. It was also imagined that the tail of the wolf contained a hair, which acted as a love-philtre and excited the tender passion. The myth of Romulus and Remus having been suckled by a wolf, arose from the simple circumstance of their nurse having been named Lupa—an explanation which sadly does away with the garland of romance that so long surrounded the story of the founders of Rome. The figure of the wolf at one time formed a standard for the Roman legions, as saith Pliny, "Caius Marius, in his second consulship, ordained that the legions of Roman soldiers only should have the eagle for their standard, and no other signe, for before time the egle marched formost indeed, but in a ranke of foure others, to wit, wolves, minotaures, horses, and bores."

The dried snout of a wolf held, in the estimation of the ancients, the same rank that a horse-shoe does now with the credulous. It was nailed upon the gates of country farms, as a counter-charm against the evil eye, and was supposed to be a powerful antidote to incantations and witchcraft. New-married ladies were wont, upon their wedding-day, to anoint the side-posts of their husbands' houses with wolves' grease, to defeat all demoniac arts. These animals bore, however, but a bad character when alive; for, exclusive of their depredations, it was imagined that if horses chanced to tread in the foot-tracks of wolves, their feet were immediately benumbed; but Pliny also says, "Verily, the great master teeth and grinders of a wolf being hanged about an horse necke, cause him that he shall never tire and be weary, be he put to never so much running in any race whatsoever."* When a territory was much infested with wolves, the following ceremony was performed with much solemnity and deep subsequent carousal:—A wolf would be caught alive, and

* Holland's Plinie's Natural Historie, ed. 1635.

his legs carefully broken. He was then dragged round the confines of the farm, being bled with a knife from time to time, so that the blood might sprinkle the ground. Being generally dead when the journey had been completed, he was buried in the very spot whence he had started on his painful race.

There was scarcely a filthy thing which the ancients did not in some way use medicinally; and we find Paulus Ægineta recommends the dried and pounded liver of a wolf, steeped in sweet wine, as a sovereign remedy for diseases of the liver, etc.

Our English word *wolf* is derived from the Saxon *wulf*, and from the same root, the German *wolf*, the Swedish *ulf*, and Danish *ulv* are probably derived. Wolves were at one time a great scourge to this country, the dense forests which formerly covered the land favouring their safety and their increase. Edgar applied himself seriously to rid his subjects of this pest, by commuting the punishments of certain crimes into the acceptance of a number of wolves' tongues from each criminal; and in Wales by commuting a tax of gold and silver imposed on the Princes of Cambria by Ethelstan, into an annual tribute of three hundred wolves' heads, which Jenaf, Prince of North Wales, paid so punctually, that by the fourth year the breed was extinct. Not so, however, in England, for like ill weeds they increased and multiplied here, rendering necessary the appointment, in the reign of the first Edward, of a *wolf-hunter* general, in the person of one Peter Corbet; and his Majesty thought it not beneath his dignity to issue a mandamus, bearing date May 14th, 1281, to all bailiffs, etc., to aid and assist the said Peter in the destruction of wolves in the counties of Gloucester, Worcester, Hereford, Shropshire, and Stafford; and Camden informs us that in Derby, lands were held at Wormhill by the duty of hunting and taking the wolves that infested that county. In the reign of Athelstan, these pests had so abounded in Yorkshire, that a retreat was built at Flixton in that county, "to defend passengers from the wolves that they should not be devoured by them." Our Saxon ancestors also called January, when wolves pair, *wolf-moneth*; and an outlaw was termed *wolfshed*, being out of the protection of the law, and as liable to be killed as that destructive beast.

A curious notice of the existence of wolves and foxes in Scotland is afforded in Bellenden's translation of Boetius.* "The wolffis are right noisome to tame beastial in all parts of Scotland, except one part thereof, named Glenmorris, in which the tame beastial gets little damage of wild beastial, especially of tods (foxes); for each house nurses a young tod certain days, and mengis (mixes) the flesh thereof,

* Edit. Edin. 1541, quoted from Magazine of Natural History.

after it be slain, with such meat as they give to their fowls or other small beasts, and so many as eat of this meat are preserved two months after from any damage of tods; for tods will eat no flesh that gusts of their own kind." The last wolf killed in Scotland is said to have fallen by the hand of Sir Ewen Cameron, about 1680; and, singular to say, the skin of this venerable quadruped may yet be in existence. In a catalogue of Mr. Donovan's sale of the London Museum, in April, 1818, there occurs the following item:—" Lot 832. Wolf, a noble animal, in a large glass case. The last wolf killed in Scotland, by Sir E. Cameron." It would be interesting to know what became of this lot.

The pairing time is January, when, after many battles with rivals, the strongest males attach themselves to the females. The female wolf prepares a warm nest for her young, of soft moss and her own hair, carefully blended together. The cubs are watched by the parents with tender solicitude, are gradually accustomed to flesh, and when sufficiently strong their education begins, and they are taken to join in the chase: not the least curious part is the discipline by which they are inured to suffering, and taught to bear pain without complaint. Their parents are said to bite, maltreat, and drag them by the tail, punishing them, if they utter a cry, until they have learned to be mute. To this quality Macaulay alludes when speaking of a wolf in his "Prophecy of Capys:"—

> "When all the pack, loud baying,
> Her bloody lair surrounds,
> She dies in silence, biting hard,
> Amidst the dying hounds."

It is curious to observe the cunning acquired by wolves in well inhabited districts, where they are eagerly sought for destruction; they then never quit cover to windward; they trot along just within the edges of the wood until they meet the wind from the open country, and are assured by their keen scent that no danger awaits them in that quarter—then they advance, keeping under cover of hedgerows as much as possible, moving in single file and treading in each other's track; narrow roads they bound across, without leaving a footprint. When a wolf contemplates a visit to a farm-yard, he first carefully reconnoitres the ground, listening, snuffing up the air, and smelling the earth; he then springs over the threshold without touching it, and seizes on his prey. In retreat his head is low, turned obliquely, with one ear forward, the other back, and the eyes glaring. He trots crouching, his brush obliterating the track of his feet till at some distance from the scene of his depredation, then feeling himself secure, he waves his tail erect in triumph, and boldly pushes on to cover.

In Northern India, wolves, together with jackals and pariah dogs, prowl about the dwellings of Europeans. Colonel Hamilton Smith relates a curious accident which befell a servant who was sleeping in a verandah with his head near the outer lattice: a wolf thrust his jaws between the bamboo, seized the man by the head, and endeavoured to drag him through; the man's shrieks awakened the whole neighbourhood, and assistance came, but though the wolf was struck at by many, he escaped. Wolves have even been known to attack sentries when single, as in the last campaign of the French armies in the vicinity of Vienna, when several of the videttes were carried off by them. During the retreat of Napoleon's army from Russia, wolves of the Siberian race followed the troops to the borders of the Rhine; specimens of these wolves shot in the vicinity, and easily distinguishable from the native breed, are still preserved in the museums of Neuwied, Frankfort, and Cassel.

Captain Lyon[*] relates the following singular instance of the cunning of a wolf which had been caught in a trap, and, being to all appearance dead, was dragged on board ship:—"The eyes, however, were observed to wink whenever an object was placed near them; some precautions were, therefore, considered necessary, and the legs being tied, the animal was hoisted

[*] Private Journal of Captain G. F. Lyon, 1824.

up with his head downwards. He then, to our surprise, made a vigorous spring at those near him, and afterwards repeatedly turned himself upwards so as to reach the rope by which he was suspended, endeavouring to gnaw it asunder, and making angry snaps at the persons who prevented him. Several heavy blows were struck on the back of his neck, and a bayonet was thrust through him, yet above a quarter of an hour elapsed before he died."

Hearne, in his journey to the Northern Ocean, says, that the wolves always burrow underground at the breeding season; and though it is natural to suppose them very fierce at those times, yet he has frequently seen the Indians go to their dens, take out the cubs, and play with them. These they never hurt, and always scrupulously put them in the den again, although they occasionally painted their faces with vermilion and red ochre, in strange and grotesque patterns.

This statement is supported by incidents which have occurred in this metropolis; there was a bitch-wolf in the Tower Menagerie, which, though excessively fond of her cubs, suffered the keepers to handle them, and even remove them from the den without evincing the slightest symptom either of anger or alarm: and a still more remarkable instance is related from observation by Mr. Bell:—"There was a wolf at the Zoological Gardens (says that able naturalist) which would always come to the front bars of the den as soon as I or any other person whom she knew, approached; she had pups, too, and so eager, in fact, was she that her little ones should share with her in the notice of her friends, that she killed all of them in succession by rubbing them against the bars of her den as she brought them forward to be fondled."

During 1850, 8807 wolves' skins were imported by the Hudson's Bay Company from their settlements; of which 8784 came from the York Fort and Mackenzie River stations; we recently had the opportunity of examining the stock, and found it principally composed of white wolves' skins from the Churchill River, with black and grey skins of every shade. The most valuable are from animals killed in the depth of winter, and of these, the white skins, which are beautifully soft and fine, are worth about thirty shillings apiece, and are exported to Hungary, where they are in great favour with the nobles as trimming for pelisses and hussar jackets; the grey wolves' skins are worth from three shillings and sixpence upwards, and are principally exported to America and the North of Europe, to be used as cloak linings.

The wolf will breed with the dog; the first instance in this country took place in 1766, when a litter, the offspring of a wolf and Pomeranian bitch, was born at Mr. Brooke's, a dealer in animals in the New-road:

one of these pups was presented to the celebrated John Hunter, who says, "Its actions were not truly those of a dog, having more quickness of attention to what passed, being more easily startled, as if particularly apprehensive of danger, quicker in transition from one action to another, being not so ready to the call, and less docile. From these peculiarities it lost its life, having been stoned to death in the streets for a mad dog."* Another of these puppies subsequently bred with other dogs, and it is a descendant of hers which lies buried in the gardens of Wilton House, and is commemorated by the following inscription on the stone which covers her—

"Here lies Lupa,
Whose grandmother was a wolf,
Whose father and grandfather were dogs, and whose
Mother was half wolf and half dog. She died
On the 16th of October, 1782,
Aged 12 years."

In another instance, where a bitch-wolf bred with a dog, two of the puppies had large black spots on a white ground; another was black, and the fourth a kind of dun. In reference to this subject, it has been well remarked by Professor Owen :†—"From the known disposition of varieties to revert to the original, it might have been expected, on the supposition that the wolf is the original of the dog, that the pro-

* Hunter's 'Animal Œconomy,' p. 320.   † Id. p. 323.

duce of the wolf and dog ought rather to have resembled the supposed original than the variety. In a litter lately obtained at the Royal Menagerie at Berlin, from a white pointer and a wolf, two of the cubs resembled the common wolf-dog, but the third was like a pointer with hanging ears."

Colonel H. Smith mentions a curious instance of the treacherous ferocity of the wolf. A butcher at New York had brought up, and believed he had tamed, a wolf, which he kept for above two years chained up in the slaughter-house, where it lived in a complete superabundance of blood and offal. One night, having occasion for some implement which he believed was accessible in the dark, he went into this little Smithfield without thinking of the wolf. He was clad in a thick frieze coat, and while stooping to grope for what he wanted, he heard the chain rattle, and in a moment was struck down by the animal springing upon him. Fortunately, a favourite cattle-dog had accompanied his master, and rushed forward to defend him: the wolf had hold of the man's collar, and being obliged to turn in his own defence, the butcher had time to draw a large knife, with which he ripped his assailant open. The same able writer relates an incident which occurred to an English gentleman, holding a high public situation in the peninsula, during a wolf-hunt in the mountains near Madrid. The sportsmen were placed in ambush, and the country-people drove the game towards them: presently an animal came bounding upward toward this gentleman, so large that he took it, while driving through the high grass and bushes, for a donkey; it was a wolf, however, whose glaring eyes meant mischief, but scared by the click of the rifle, he turned and made his escape, though a bullet whistled after him; at the close of the hunt seven were found slain, and so large were they that this gentleman, though of uncommon strength, could not lift one entirely from the ground.

The wolf of America is at times remarkable for cowardice, though bold enough when pressed by hunger, or with other wolves. Mr. R. C. Taylor, of Philadelphia, states that this animal, when trapped, is silent, subdued, and unresisting. He was present when a fine young wolf, about fifteen months old, was taken by surprise, and suddenly attacked with a club. The animal offered no resistance, but, crouching down in the supplicating manner of a dog, suffered himself to be knocked on the head. An old hunter told Mr. Taylor that he had frequently taken a wolf out of the trap, and compelled it by a few blows to lie down by his side, while he reset his trap.

The Esquimaux wolf-trap is made of strong slabs of ice, long and so narrow, that a fox can with difficulty turn himself in it, and a wolf must actually remain in the position in which he is taken. The door is a heavy portcullis of ice, sliding in two well-secured grooves of the same substance, and is kept up by a line which, passing over the top of the trap, is carried through a hole at the furthest extremity. To the end of the line is fastened a small hoop of whalebone, and to this any kind of flesh bait is attached. From the slab which terminates the trap, a projection of ice, or a peg of bone or wood, points inwards near the bottom, and under this the hoop is slightly hooked; the slightest pull at the bait liberates it, the door falls in an instant, and the wolf is speared where he lies.

Sir John Richardson states that, when near the Copper Mine River in North America, he had more than once an opportunity of seeing a single wolf in pursuit of a reindeer, and especially on Point Lake, when covered with ice, when a fine buck reindeer was overtaken by a large white wolf, and disabled by a bite in the flank. An Indian, who was concealed, ran in and cut the deer's throat with his knife, the wolf at once relinquishing his prey and sneaking off. In the chase the poor deer urged its flight by great bounds, which for a time exceeded the speed of the wolf; but it stopped so frequently to gaze on its relentless enemy, that the latter, toiling on at a long gallop (so admirably described by Byron), with his

tongue lolling out of his mouth, gradually came up.
After each hasty look, the deer redoubled its efforts
to escape, but either exhausted by fatigue, or enervated
by fear, it became, just before it was overtaken, scarcely
able to keep its feet.

Captain Lyon gives some interesting illustrations
of the habits of the wolves of Melville Peninsula,
which were sadly destructive to his dogs. "A fine
dog was lost in the afternoon. It had strayed to the
hummocks ahead, without its master, and Mr. Elder,
who was near the spot, saw five wolves rush at, attack,
and devour it, in an incredibly short space of time;
before he could reach the place, the carcase was torn
in pieces, and he found only the lower part of one leg.
The boldness of the wolves was altogether astonishing,
as they were almost constantly seen among the hummocks, or lying quiet at no great distance in wait for
the dogs. From all we observed, I have no reason to
suppose that they would attack a single unarmed man,
both English and Esquimaux frequently passing them
without a stick in their hands. The animals, however,
exhibited no symptoms of fear, but rather a kind of
tacit agreement not to be the beginners of a quarrel,
even though they might have been certain of proving
victorious."* Another time, when pressed by hunger,
the wolves broke into a snow-hut, in which were a

* Private Journal of Captain G. F. Lyon, 1824.

couple of newly-purchased Esquimaux dogs, and carried the poor animals off, but not without some difficulty, for even the ceiling of the hut was next morning found sprinkled with blood and hair. When the alarm was given and the wolves were fired at, one of them was observed carrying a dead dog in his mouth, clear of the ground, and going with ease at a canter, notwithstanding the animal was of his own weight. It was curious to observe the fear these dogs seemed at times to entertain of wolves.

During Sir John Richardson's residence at Cumberland House in 1820, a wolf which had been prowling round the fort, was wounded by a musket-ball, and driven off, but returned after dark, whilst the blood was still flowing from its wound, and carried off a dog from amongst fifty others, but which had not the courage to unite in an attack on their enemy. The same writer says that he has frequently observed an Indian dog, after being worsted in combat with a black wolf, retreat into a corner and howl at intervals for an hour together. These Indian dogs also howl piteously when apprehensive of punishment, and throw themselves into attitudes strongly resembling those of a wolf when caught in a trap.

Foxes are frequently taken in the pitfalls set for wolves, and seem to possess more cunning. An odd incident is related by Mr. Lloyd. A fox was lying at the bottom of a pitfall, apparently helpless, when a very stout peasant, having placed a ladder, began to descend with cautious and creaking steps to destroy the vermin. Reynard, however, thought he might benefit by the ladder as well as his corpulent visitor, and just as the latter reached the ground, jumped first on his stern, then on his shoulder, skipped out of the pit, and was off in a moment, leaving the man staring and swearing at his impudent escape.

Captain Lyon mentions an instance of the sagacity of the fox; he had caught and tamed one of these animals, which he kept on deck in a small hutch with a scope of chain. Finding himself repeatedly drawn out of his hutch by this, the sagacious little fellow, whenever he retreated within his castle, took the chain in his mouth, and drew it so completely in after him that no one, who valued his fingers, would endeavour to take hold of the end attached to the staple.

Mr. Lloyd mentions a curious contest that took place in the vicinity of Uddeholm. A peasant had just got into bed when his ears were assailed by a tremendous uproar in his cattle-shed. On hearing this noise he jumped up, and though almost in a state of nudity, rushed into the building to see what was the matter; here he found an immense wolf, which he gallantly seized by the ears, and called out most lustily for assistance. His wife, the gallant Trulla, came to his aid, armed with a hatchet, with which she severely wounded the wolf's head, but it was not until she had driven the handle of the hatchet down the animal's throat, that she succeeded in despatching him; during the conflict the man's hands and wrists were bitten through and through, and, when seen by Mr. Lloyd, the wounds were not healed.

Like dogs, wolves are capable of strong attachment; but such instances are comparatively rare; the most striking, perhaps, was that recorded by M. Frederic Cuvier, as having come under his notice at the Ménagerie du Roi at Paris. The wolf in question was brought up as a young dog, became familiar with persons he was in the habit of seeing, and in particular followed his master everywhere, evincing chagrin at his absence, obeying his voice, and showing a degree of submission scarcely differing in any respect from that of the most thoroughly domesticated dog. His master, being obliged to be absent for a time, presented his pet to the menagerie, where he was confined in a den. Here he became disconsolate, pined, and would scarcely take food; at length he was reconciled to his new situation, recovered his health, became attached to his keepers, and appeared to have forgotten 'auld lang syne,' when, after the lapse of eighteen months, his old master returned. At the first sound of his voice—that well-known, much-loved voice—the wolf, which had not perceived him in a crowd of persons, exhibited the most lively joy, and being set at liberty, lavished upon him the most affectionate caresses, just as the most attached dog would have done. With some difficulty he was enticed to his den. But a second separation was followed by similiar demonstrations of sorrow to the former, which, however, again yielded to time. Three years passed away, and the wolf was living happily with a dog which had been placed with him, when his master again appeared, and again the long-lost but well-remembered voice was instantly replied to by the most impatient cries, redoubled as soon as the poor fellow was at liberty; rushing to his master, he placed his fore-feet on his shoulders, licking his face with every mark of the most lively joy, and menacing the keepers who offered to remove him. A third separation, however, took place, but it was too much for the poor creature's temper; he became gloomy, refused his food, and for some time it was feared he would die. Time, however, which blunts the grief of wolves as well as of men, brought comfort to his wounded heart, and his health gradually returned; but, looking upon mankind as false deceivers, he no longer permitted the caresses of any but his keepers, manifesting to all strangers the savageness and moroseness of his species.

Another instance of the attachment of wolves is

mentioned by Mr. Lloyd in his work on the Sports of the North, from which we have frequently quoted. Mr. Greiff, who had studied the habits of wild animals, for which his position as *öfver jäg mästare* afforded peculiar facilities, says:—"I reared up two young wolves until they were full-grown. They were male and female. The latter became so tame that she played with me and licked my hands, and I had her often with me in the sledge in winter. Once when I was absent, she got loose from the chain, and was away three days. When I returned home I went out on a hill and called, 'Where's my Tussa?' as she was named, when she immediately came home, and fondled with me like the most friendly dog."

Between the dog and the wolf there is a natural enmity, and those animals seldom encounter each other on at all equal terms without a combat taking place. Should the wolf prove victorious, he devours his adversary, but if the contrary be the case, the dog leaves untouched the carcase of his antagonist.

The wolf feeds on the rat, hare, fox, badger, roebuck, stag, reindeer, and elk; likewise upon blackcock and capercali. He is possessed of great strength, especially in the muscles of the neck and jaws, is said always to seize his prey by the throat, and when it happens to be a large animal, as the elk, he is ofter dragged for a considerable distance.

After a deep fall of snow the wolf is unusually ferocious; if he besmears himself with the blood of a victim, or is so wounded that blood flows, it is positively asserted that his companions will instantly kill and devour him.

In the year 1799 a peasant at Frederickshall, in Norway, was looking out of his cottage window, when he espied a large wolf enter his premises and seize one of his goats. At this time he had a child of eighteen months old in his arms; he incautiously laid her down in a small porch fronting the house, and, catching hold of a stick, the nearest weapon at hand, attacked the wolf which was in the act of carrying off the goat. The wolf dropped this, and getting sight of the child, in the twinkling of an eye seized it, threw it across his shoulders, and was off like lightning. He made good his escape, and not a vestige was ever seen of the child.

Wolves are found all over Scandinavia, but are most common in the midland and northern provinces of Sweden. Like "Elia," they are very partial to young pig, a failing taken advantage of by sportsmen thus: they sew up in a sack a small porker, leaving only his snout free, and place him in a sledge, to the back of which is fastened by a rope about fifty feet long, a small bundle of straw, covered with black sheepskin; this, when the sledge is in motion, dangles about like a young pig.

During a very severe winter a party started in the vicinity of Forsbacka, well provided with guns, etc. On reaching a likely spot they pinched the pig, which squealed lustily, and, as they anticipated, soon drew a multitude of famished wolves about the sledge. When these had approached within range, the party opened fire on them, and shot several; all that were either killed or wounded were quickly torn to pieces and devoured by their companions; but the blood with which the ravenous beasts had now glutted themselves, only served to make them more savage than before, and, in spite of the fire kept up by the party, they advanced close to the sledge, apparently determined on making an instant attack. To preserve the party, therefore, the pig was thrown to the wolves, which had for a moment the effect of diverting their attention. Whilst this was going forward, the horse, driven to desperation by the near approach of the wolves, struggled and plunged so violently that he broke the shafts to pieces, galloped off, and made good his escape. The pig was devoured, and the wolves again threatened to attack the sportsmen. The captain and his friends, finding matters had become serious, turned the sledge bottom up, and took shelter beneath it, in which position they remained many hours, the wolves making repeated attempts to get at them by tearing the sledge with their teeth; but at length the party were relieved by friends from their perilous position.

Lieutenant Oldenburg once witnessed a curious occurrence. He was standing near the margin of a large lake which at that time was frozen over. At some little distance from the land a small aperture had been made for the purpose of procuring water, and at this hole a pig was drinking. Whilst looking towards the horizon, the Lieutenant saw a mere speck or ball, as it were, rapidly moving along the ice: presently this took the form of a large wolf, which was making for the pig at top speed. Lieutenant Oldenburg now seized his gun, and ran to the assistance of the pig; but before he got up to the spot, the wolf had closed with the porker, which, though of large size, he tumbled over and over in a trice. His attention was so much occupied, that Lieutenant Oldenburg was able to approach within a few paces and despatch him with a shot. A piece as large as a man's foot had been torn out of the pig's hind-quarters; and he was so terribly frightened that he followed the Lieutenant home like a dog, and would not quit his heels for a moment.

Mr. Lloyd mentions an incident that befell him in consequence of swine mistaking his dogs for wolves, to which they bear the most instinctive antipathy. One day, in the depth of winter, accompanied by his Irish servant, he struck into the forest, in the vicinity of Carlstadt, for the purpose of shooting capercali. Towards evening they came to a small hamlet, situated in the recesses of the forest. Here an old sow with her litter were feeding; and immediately on seeing the two valuable pointers which accompanied the sportsman, she made a determined and most ferocious dash at them. The servant had a light spear in his hand, similar to that used by our lancers. This Mr. Lloyd seized, and directing Paddy to throw the dogs over a fence, received the charge of the pig with a heavy blow across the snout with the butt-end of the spear. Nothing daunted, she made her next attack upon him; and, in self-defence, he was obliged to give her a home thrust with the blade of the spear. These attacks she repeated three several times, always getting the spear up to the hilt in her head or neck. Then, and not before, did she slowly retreat, bleeding at all points. The peasants, supposing Mr. Lloyd to be the aggressor, assumed a very hostile aspect, and it was only by showing a bold bearing, and menacing them with his gun, that he escaped in safety.

A poor soldier was one day, in the depth of winter, crossing the large lake called Storsyön, and was attacked by a drove of wolves. His only weapon was a sword, with which he defended himself so gallantly, that he killed and wounded several wolves, and succeeded in driving off the remainder. After a time, he was again attacked by the same drove, but was now unable to extricate himself from his perilous situation in the same manner as before; for having neglected to wipe the blood from his sword after the former encounter, it had become firmly frozen to the scabbard. The ferocious beasts, therefore, quickly closed with him, killed and devoured him. If we remember aright, Sir John Kincaid, the present gallant Exon of the Yeoman Guard, nearly lost his life at Waterloo, from a somewhat similar cause. He had been skirmishing all the earlier part of the day with the Rifles, when a sudden charge of French cavalry placed him in a great danger. He essayed to draw his sabre, tugged and tugged, but the trusty steel had become firmly rusted to the scabbard; and we believe that he owed his life to an accidental diversion of the attention of the attacking troopers.

Closely resembling in many respects the wolf, the Jackal is widely spread over India, Asia, and Africa. These animals hunt in packs, and there are few sounds more startling to the unaccustomed ear than a chorus of their cries. "We hardly know," says Captain Beechey, "a sound which partakes less of harmony than that which is at present in question; and indeed the sudden burst of the answering long, protracted scream, succeeding immediately to the opening note, is scarcely less impressive than the roll of the thunder clap immediately after a flash of lightning. The effect of this music is very much increased when the first note is heard in the distance, a circumstance which often occurs, and the answering yell bursts out from several points at once, within a few yards or feet of the place where the auditors are sleeping."

Poultry and the smaller animals, together with dead bodies, are the ordinary food of jackals, but when rendered bold by hunger, they will occasionally attack the larger quadrupeds and even man.

A bold, undaunted presence and defiant aspect, generally proves the best protection when an unarmed man is threatened by these or other animals, but artifice is sometimes necessary. A ludicrous instance is related by an old quartermaster (whom we knew some years ago), in a small volume of memoirs.* At Christmas, 1826, he was sent up the country to a mission, about thirty-two miles from San Francisco. He and the others erected a tent: after which they all lay down on the ground. "I slept like a top," says he, "till four the next morning, at which time I was awakened by the man whose duty it was to officiate as cook for the day, who told me if I would go up to the village and get a light, he would have a good breakfast ready for the lads by the time they awoke. I must describe my dress, for that very dress saved my life. Over the rest of my clothing, as a seaman, I had a huge frock made from the skin of a

* 'Thirty-six Years of a Seafaring Life:' 1839.

rein-deer. It was long enough, when let down, to cover my feet well, and turned up at foot, buttoning all round the skirt. At the top was a hood, made from the skin taken off the head of a bear, ears and all. In front was a square lappel, which, in the day, hung loosely over the breast, but at night, buttoned just behind the ears, leaving only the mouth, nose, and eyes free for respiration, so that one, with such a dress, might lie down anywhere and sleep, warm and comfortable. Mr. S—— had given eight dollars for it in Kamtchatka, and, on our return to more genial climes, forgot the future, and gave it to me. Fancy, then, my figure thus accoutred, issuing from under the canvas vent, with a lantern in my hand. I had not advanced twenty yards, when first only two or three, and then an immense number of jackals surrounded me. I was at first disposed to think but lightly of them: but seeing their numbers increase so rapidly, I grew alarmed, and probably gave way to fear sooner than I ought. A few shots from the tent would probably have sent them away with speed, but no one saw me. Every moment they drew closer and closer in a complete round, and seemed to look at me with determined hunger. For some moments I remained in a most dreadful state of alarm. It just then occurred to me that I once heard of a boy who **had driven back a** bull out of a field by walking backwards on his hands and feet. Fortunate thought! I caught at the idea; in a moment I was on all-fours, with my head as near the earth as I could keep it, and commenced cutting all the capers of which I was capable. The jackals, who no doubt had never seen so strange an animal, first stopped, then retreated, and, as I drew near the tent, flew in all directions. The men awoke just in time to see my danger, and have a hearty laugh at me and the jackals."

Our old friend was more fortunate than a certain youth who attempted to rob an orchard by deluding a fierce bulldog with this approach *à posteriori*, but who, to his sorrow, found the dog too knowing, for he carried to his dying day the marks of the guardian's teeth in that spot where honour has its seat.

The same quartermaster told us a quaint story of a fright another of the crew received from these jackals.

"Whilst at San Francisco the ship's crew were laying in a store of provisions; a large tent was erected on shore for salting the meat; the cooper lived in it, and hung up his hammock at one end. The beef which had been killed during the day was also hung up all around, in readiness for salting. One night a large pack of jackals came down from the woods, and being attracted by the smell of the meat, soon got into the tent, and pulling at one of the sides of beef, brought it down with a crash, which woke the old cooper, who was a remarkably stout, and rather nervous man. Finding himself thus surrounded in the dead of the night by wild-beasts, whose forms and size, dimly seen, were magnified by his fears, he fired off his musket, and clasping his arms, in an agony of terror, round a quarter of beef which hung close to his hammock, was found perfectly senseless by an officer who came to see the cause of the alarm. Some difficulty was experienced in getting him to relinquish his hold of the beef—which he stuck to like a Briton —and it was several days before his nerves recovered from the shock of the fright.

The wolf and jackal tribes are by no means without their use in the economy of nature, though from their predatory habits they are justly regarded as pests in the countries they infest: that they will disturb the dead and rifle the graves is true, but they also clear away offal, and, with vultures, are the scavengers of hot countries; they follow on the track of herds, and put a speedy end to the weak, the wounded, and the dying; they are the most useful, though most disgusting of camp-followers, and after a battle, when thousands of corpses of men and horses are collected within a limited space, they are of essential service.

> "I stood in a swampy field of battle,
> With bones and skulls I made a rattle
> To frighten the wolf and carrion crow
> And the homeless dog—but they would not go

> So off I flew—for how could I bear
> To see them gorge their dainty fare?"
>
> COLERIDGE.

Revolting and heart-sickening though such scenes may be, the evil is less than would result from the undisturbed decay of the dead; were that to take place, the air would hang heavy with pestilence, and the winds of heaven laden with noisome exhalations would carry death and desolation far and near, rendering still more terrible the horrors and calamities of war.

CHAPTER V.

ANTIQUITY OF HORSEMANSHIP.—BUCEPHALUS.—BRITISH WAR-CHARIOTS.—ABORIGINAL PONIES.—STONE HORSE-COLLARS.—FIRST RACERS.—MASTER OF HORSE.—ARABIANS.—LORD BURGHLEY. —LORD HERBERT OF CHERBURY.—SPEED OF HORSES.—GODOLPHIN ARABIAN.—EAGERNESS OF HORSES.—AN OLD WARRIOR.—A SURPRISE.—FIRST HORSES IN AMERICA.—A DOMIDOR.—BREAKING A HORSE.—SOUTH AMERICAN STEEDS.—TURNING THE TABLES.— SWIMMING.—TURKOMAN HORSES.—A HEROINE.—ABD-EL-KADIR. —ARAB MAXIMS.—THE WARRIORS OF THE DESERT.—A FIGHT.— THE DEFEAT.—A COMPLIMENTARY EXCUSE.—A NICE DISTINCTION. —MUNGO PARK AND HIS HORSE.—A SAD LOSS.—A NOBLE ANIMAL.—ROBBERS OF THE DESERT.—A PET.—THE CAMANCHEES.— "SMOKING" HORSES.—"CHARLEY."—THE FAITHFUL STEED.— WILD HORSES.—A BOLD STROKE.—A GALLANT CHARGER.—CORUNNA.—NEBUCHADNEZZAR.

"Hast thou given the horse strength? Hast thou clothed his neck with thunder? Canst thou make him afraid as a grasshopper? The glory of his nostrils is terrible.

"He paweth in the valley, and rejoiceth in his strength. He goeth on to meet the armed men. He mocketh at fear, and is not affrighted; neither turneth he back from the sword.

"He swalloweth the ground with fierceness and rage, neither believeth he that it is the sound of the trumpet.

"He saith among the trumpets, ha! ha! and he smelleth the battle afar off, the thunder of the captains and the shouting."

JOB xxxix.

WITH the exception of Genesis, the book of Job is,

we believe, considered to be the most ancient writing in the world; and it is interesting to remark, that in the above spirited and admirable description the horse is spoken of at that early period—a period antecedent to Abraham—as trained to battle, and familiar with war. In all probability, the use of the horse in warfare is almost coeval with war itself; and from a verse in the same chapter as the above, it is clear that the horse was employed then, as now, in the chase of the ostrich, of which bird it is said, "What time she lifteth up herself on high, she scorneth the horse and his rider." But a still earlier intimation of the horse being subdued by man, is conveyed in the seventeenth verse of the forty-ninth chapter of Genesis:—"An adder in the path, that biteth the horse's heels, so that *his rider* falleth backward." From motives which are matter of speculation, horses were not permitted to be bred by the people of Israel, nor were they permitted to use them. Indeed, it was not till the time of Solomon, five hundred years after the Israelites had left Egypt, that the horse was domesticated among them. It is curious to know that the price of horses is stated to have been 150 shekels of silver, or rather more than £17 each,* for which sum they were obtained from Egypt.

It is considered that the first domestication of the

* 1 Kings x. 29.

horse took place in Central Asia, whence the knowledge of his usefulness radiated to China, India, and Egypt, and it was most probably in ancient Egypt that systematic attention was first paid to improving the breed of these animals; for there are abundant pictorial and carved representations of steeds whose symmetry and beauty attest that they were designed from high-bred types. It was in High Asia that the bridle, the true saddle, the stirrup, and probably the horseshoe were invented, and with many of those nations a horse, a mare, and a colt were fixed nominal standards of value.

Bucephalus, probably the most celebrated horse in the world, was bought for sixteen talents from Philonicus out of his breeding pastures of Pharsalia, and it is known that he was a *skewbald*, that is, white clouded with large deep-bay spots: this particular breed was valued by the Parthians above all others, but by the Romans it was disliked, because easily seen in the dark. Bucephalus was ridden by Alexander at the battle of the Hydaspes, and there received his death-wound. Disobedient for once to the command of his master, he galloped from the heat of the fight, brought Alexander to a place where he was secure from danger, knelt (as was his custom) for him to alight, and having thus like a true and faithful servant discharged his duty to the last, he trembled, dropped down, and died:

> "Master, go on, and I will follow thee
> To the last gasp, with love and loyalty."

In Revelations, Triumph, War, Pestilence, and Death, are respectively typified by a white, a red, a black, and a pale horse; and in Europe the black horse was long considered as the form of an evil demon. Curious enough, among the modern pagan Asiatics, Schaman sorcery is usually performed with images of small horses suspended from a rope; and a sort of idolatrous worship is admitted even by Mohammedans, when effigies of the horse of Hoscin, or of that of Khizr, the St. George of Islam, are produced.

The thirty-third chapter of the fourth book of Cæsar's Commentaries has especial interest as detailing accurately the mode of equestrian warfare of the aborigines of Britain:—"Their mode of fighting with their chariots is this: firstly, they drive about in all directions and throw their weapons, and generally break the ranks of the enemy with the very dread of their horses and the noise of their wheels; and when they have worked themselves in between the troops of horse, leap from their chariots, and engage on foot. The charioteers meantime withdraw some little distance from the battle, and so place themselves with the chariots, that if their masters are overpowered by the number of the enemy, they may have a ready retreat to their own troops. Thus they display in battle

the speed of horse with the firmness of infantry; and
by daily practice and exercise attain to such expertness, that they are accustomed even on a declining and
steep place to check their horses at full speed, and
manage and turn them in an instant, and run along
the pole and stand on the yoke, and thence betake
themselves with the greatest celerity to their chariots
again." The particular description of horse here alluded to is uncertain, but there was then in these
islands a race of indigenous ponies which is still represented by the Shetland, Welsh, New Forest, and
Dartmoor breeds: their stature is attested by a remark
of St. Austin:—" The *mannii*, or ponies brought from
Britain, were chiefly in use among strolling performers
to exhibit in feats of their craft;" and it was the fashion at that time to shave all the upper parts of the
shaggy bodies of these ponies in summer, somewhat
after the fashion of the *clippers* of the present day.

The county Argyle in Scotland is said to derive its
name from *Are-Gael*,—the breeding or horse stud of
the Gael; and in a superb work recently published,
called the 'Archæology of Scotland,' there is a description of a truly remarkable discovery, throwing
light on the charioteering of the Celts. There have
been dug up, near the parallel roads of Glen Roy, two
stone horse-collars, the one formed of trap or whinstone, the other of a fine-grained red granite: these

bear all the evidence of first-rate workmanship, are highly polished, and are of the full size of a collar adapted to a small highland horse, bearing a close imitation of the details of a horse-collar of common materials in the folds of the leather, the nails, buckles, etc. It has been suggested by antiquarians, that the amphitheatre of Glen Roy might have been the scene of ancient public games, and that these stone collars might be intended to commemorate the victor in the race.

Hengist, the name of the founder of the Saxon dynasty, signified an entire horse; and by the Saxons the horse was an object of superstitious veneration. Of this there remains an example which must be familiar to all who in the old coaching days rode through White Horse Vale in Berkshire. The turf on the side of a hill has been cut away, displaying the chalk beneath in the figure of a gigantic horse, covering many hundred square feet. This is a genuine Saxon relic, and has been preserved by a day being annually kept as high festival, on which all weeds are carefully cleared from the figure, and the outline restored.*

The Anglo-Saxons are supposed to have first used the horse in ploughing, about the latter part of the tenth century; on the border of the Bayeux tapestry, representing the landing of William the Conqueror

* See 'The Scouring of the White Horse.'

and the Battle of Hastings (A. D. 1066), there is a representation of a man driving a harrow, the earliest instance we believe of horses thus used in field labour.

Horse-racing was introduced into Britain A.D. 930, when Hugh the Great, head of the house of Capet, monarchs of France, presented to Athelstan, whose sister Edelswitha he wooed and won, several running horses (*equos cursores* of the old Chronicle) magnificently caparisoned. Athelstan seems to have attached due importance to this improvement upon the previous breed, since he issued a decree prohibiting the exportation of horses without his license. The most marked improvement, however, took place at the Norman Conquest, the martial barons bringing with them a large force of cavalry, and it was, by the way, to their superiority in that important arm that the victory of Hastings was in a great measure to be ascribed.

The Easter and Whitsuntide holidays were especially famous among our forefathers for racing, as mentioned in the old metrical romance of Sir Bevis of Southampton;—

> "In somer, at Whitsontyde,
> Whan knightes most on horseback ryde,
> A cours let they make on a daye,
> Steedes and palfraye for to assaye
> Whiche horse that best may ren.
> Three myles the cours was then:

> Who that might 'ryde hym shoulde
> Have forty pounds of redy golde."

The office of Master of the Horse dates back to Alfred the Great; the ancient Chronicles relate the attention paid by him to the breeding and improvement of the horse, to carry out which in the most efficient manner an officer was appointed, called *Hors Than*, or *Horse Thane*,—Master of the Horse; and during every succeeding reign this officer has held high rank, being near the royal person on all state occasions.

We may form some idea of the value of a horse about that time by a document bearing date A.D. 1000, which states that if a horse be destroyed or negligently lost, the compensation to be demanded was thirty shillings, for a mare or colt twenty shillings, and for a man one pound. Of course the currency of the present day differs widely from that of the above period.

The year of grace 1211 is important in equestrian annals, as witnessing the introduction into England of the first of the Arabian stock; and about the same time another was presented by Alexander I., king of Scotland, to the church of St. Andrew's, though what relation a racehorse had to the church has been a knotty point for antiquaries. Both these animals were true barbs from Morocco, procured doubtless through the agency of Jew dealers. There is no breed which has exercised so great an influence upon the stock of these islands as the Arabian, and none more deserving of kindness for the admirable qualities they possess. Kindness and forbearance towards animals is inculcated by the Koran, and it was a cutting satire upon our boasted civilization when, in allusion to this point, a Moor remarked to Colonel Hamilton Smith, "It is not in your book."

Henry VIII., with that wrongheaded obstinacy so characteristic of him, issued most arbitrary laws respecting horses, ordering all horses to be destroyed at Michaelmas in each year that were not likely to produce a valuable breed, and commanding that all his prelates and nobles, and " all those whose wives wore velvet bonnets," should keep horses for the saddle at least fifteen hands high. The effect of these miserable regulations was so injurious, that, forty-one years afterwards, Elizabeth could only muster three thousand mounted soldiers to repel the expected Armada.* Old Andrew Fuller relates a quaint anecdote of Lord Burghley, the celebrated sage councillor of Queen Bess:—" When some noblemen had got William Cecil

---

* Mr. Macculloch estimates that there are in Great Britain from 1,400,000 to 1,500,000 horses employed for various purposes of pleasure and utility: taking their average worth at from £10 to £12, their total value would be from £14,000,000 to £18,000,000 exclusive of the young horses.

to ride with them a-hunting, and the sport began to be cold, 'What call you this?' said the Treasurer.—'Oh, now the dogs are at fault,' was the reply.—'Yea,' quoth the Treasurer, 'take *me* again in such a fault, and I'll give you leave to punish me.'" Hunting was not his forte.

In the reign of James I. races were established in many parts of the kingdom; and the races were then called bell courses, the prize being a silver bell, whence the expression to "bear off the bell." In the reign of Charles I. races were held in Hyde Park and at Newmarket, and Charles II. most warmly patronized them, entering horses at Newmarket in his own name; and about this time the bells were converted into cups, or other species of plate, valued at a hundred guineas each. In those earlier days professional jockeys were unknown, but it is curious to hear the opinion of a celebrated writer and distinguished man, Lord Herbert of Cherbury. "The exercise," says he, "I do not approve of, is running of horses, there being much cheating in that kind. Neither do I see why a brave man should delight in a creature whose chief use is to help him *to run away!*" Lord Herbert might have been a great philosopher, but he certainly would not have been qualified to be a member of the Jockey Club. Cromwell, however, who had himself trained the finest regiment of cavalry then in existence, was aware of the importance of speed and bottom, and Charles II. obtained a large number of mares and stallions from the Levant, so that the Arabian blood was freely mingled with that of the native breed.

The feats of celebrated horses are duly chronicled in books devoted to sporting subjects, and we shall merely notice them incidentally. The most extraordinary instance, perhaps, of the stoutness, as well as speed of the race-horse, was afforded by Quibbler, who, in December, 1786, ran twenty-three miles round the flat at Newmarket in fifty-seven minutes and ten seconds! In 1772, a mile was run by Firetail in one minute and four seconds, and Flying Childers ran over the Beacon Course (four miles, one furlong, one hundred and thirty-eight yards) in seven minutes and thirty seconds! On the 29th of September, 1838, a trial of speed took place between the Oural Cossacks and the Kerguise Kaisaks over a course of eighteen versts, said to be equal to thirteen and a half English miles. The race was run by many horses of great speed, but gained by twins who ran neck and neck the whole distance, arriving at the winning post in twenty-four minutes thirty-five seconds! And it is said that the Sultan's son rode a Kerguise Kaisak black horse over the same course in nineteen minutes.

In 1745, Mr. Thornton rode from Stilton to London,

back, and again to London, making two hundred and fifteen miles, in eleven hours, on the turnpike road and uneven ground; and when the wretched state of the roads at that period is considered, the feat was truly remarkable.

Perhaps the most singular struggle on record was that between Tarragon, Handel, and Astbury, at Newcastle-under-Lyne. Of the first *three* heats there was no winner, Tarragon and Handel being each time nose and nose: and although Astbury was stated to have been third in the first heat, yet he was so nearly on a level with the others that there was a difficulty in placing him as such. After the second heat the steward requested two other gentlemen to look with him steadily as they came, to try to decide in favour of one of them, but it was impossible to do so. In the third dead heat Tarragon and Handel had struggled with each other until they reeled about as if they were drunk, and could scarcely carry their riders to the scales. Astbury, who had lain by after the first heat, then came out and won.

One of the most celebrated race-horses this country has seen was the Godolphin Arabian, who was bought in France when actually engaged in drawing a cart. Between this noble animal and a cat a most loving friendship existed. When in the stable puss always either sat upon his back, or nestled as closely to him as she could; and at his death she refused her food, pined away, and died. Mr. Holcroft gives a similar relation of a racer and a cat whom the horse used to take up in his mouth and mount on his back without hurting her, she perfectly understanding this singular mode of conveyance. There was another celebrated horse of yore called the *Mad Arabian*, from his great ferocity and ungovernable temper. This horse—Chillaby by name—savagely tore in pieces the figure of a man purposely placed in his way, and could only be approached by one groom. Yet with all this ferocity he evinced the most tender affection for a lamb, who used to employ himself for many an hour in butting away flies that annoyed his friend.

It is well known how thoroughly racers enter into the spirit of the course. Of this a noble horse called Forrester presented a remarkable illustration. Forrester had won many a hardly-contested race, but in an evil hour was matched against an extraordinary horse called Elephant. It was a four-mile course, and at the distance post the horses were nose to nose. Between this and the winning-post Elephant got a little ahead. Forrester made every possible effort to recover this lost ground, until, finding all his efforts ineffectual, he made one desperate plunge, seized his antagonist by the jaw, and could scarcely be forced to quit his hold. A similar incident occurred in 1753, when a fine horse belonging to Mr. Quin was rendered so frantic at finding his antagonist gradually passing him, that he seized him by the leg, and both riders were obliged to dismount and combine their efforts to separate the animals.

In battle, horses have been known to seize the opposing charger with the utmost fury, and thus to assist the sabre of his rider. This calls to our mind the death of an old war-horse at Stangleton Lodge, near Bedford. This fine old fellow had served in one of our light cavalry regiments which had played a conspicuous part at Waterloo. His hide bore the marks of several wounds by sabre and lance, and no less than eight musket-balls were found in his body! Notwithstanding this he had attained to the ripe age of twenty-seven.

The New World is indebted for the myriads of wild horses which swarm upon the Pampas of the South and the Prairies of the North, to the Spanish stock carried by Cortez to Mexico, and to Peru by Pizarro. In genial climates it was natural that with abundant herbage and few dangerous enemies, animals of such power and intelligence should increase and multiply with great rapidity. Dr. Rengger notes the first horses in Paraguay to have been imported from Spain and the Canaries in 1537, and Azara found in the Archives of Ascension, a document proving that Irala, in 1551, bought a Spanish horse for the sum of fifteen thousand florins.

According to Herrara, the Spanish historian, horses were objects of the greatest astonishment to all the people of New Spain. At first they imagined the horse and his rider, like the centaurs of the ancients, to be some monstrous animal of a terrible form, and supposing their food was that of men, brought flesh and bread to nourish them. Even after they discovered their mistake they believed the horses devoured men in battle, and when they neighed, thought they were demanding their prey. A curious incident occurred when Pizarro on one occasion was in great straits, being hemmed in by a body of ten thousand men of resolute bearing, and eager to drive the invaders into the sea. As the Spaniards were making their way, hotly pressed, one of the cavaliers was thrown from his horse. This, which at first sight might be considered an untoward event, was the salvation of the party, for the Indians were so astonished at this spontaneous separation of what they supposed to be one and the same being, that not knowing what would happen next, they actually took to flight and left the coast clear for the Spaniards to reach their ships.

The inhabitants of the Isles of Peten listened attentively to the preaching of the Franciscan Friars who accompanied the expedition of Cortez, and consented to the instant demolition of their idols and the erection of the Cross upon their ruins. How far these hurried conversions were founded on conviction is shown by the following anecdote. Cortez on his departure left among this friendly people one of his horses which had been disabled by an injury in the foot. The Indians felt a reverence for the animal as in some way connected with the mysterious power of the white men. When their visitors had gone, they offered flowers to the horse, and, as is said, prepared for him many savoury messes of poultry, such as they would have administered to their own sick. Under this extraordinary diet the poor animal pined away and died: the Indians raised his effigy in stone, and placing it in one of their temples, did homage to it as to a deity. In 1618, when two Franciscan friars came to preach the Gospel in these regions, then scarcely better known to the Spaniards than before the time of Cortez, one of the most remarkable objects which they found was this statue of a horse, receiving the homage of the Indian worshipers as the God of thunder and lightning!

The admirable skill of the South Americans as horsemen is everywhere acknowledged, and has been described by many writers; the following account, however, by Mr. Darwin, is so truthful and spirited, that it conveys the best idea of their exploits. "One evening a 'domidor' (subduer of horses) came for the purpose of breaking in some colts. I will describe the preparatory steps, for I believe they have not been mentioned by other travellers. A troop of wild young horses is driven into the corral, or large enclosure of stakes, and the door is shut. We will suppose that one man alone has to catch and mount a horse which as yet had never felt bridle or saddle. I conceive, except by a Guacho, such a feat would be utterly impracticable. The Guacho picks out a full-grown colt; and as the beast rushes round the circus, he throws his lasso so as to catch both the front legs. Instantly the horse rolls over with a heavy shock, and whilst struggling on the ground, the Guacho, holding the lasso tight, makes a circle so as to catch one of the hind legs just beneath the fetlock, and draws it close to the two front. He then hitches the lasso, so that the three legs are bound together; then sitting on the horse's neck, he fixes a strong bridle, without a bit, to the lower jaw. This he does by passing a narrow thong through the eyeholes at the end of the reins, and several times round both jaw and tongue. The two front legs are now tied closely together with a strong leathern thong fastened by a slip-knot; the lasso which bound the three together being then loosed, the horse rises with difficulty. The Guacho, now holding fast the bridle fixed to the lower jaw, leads the horse outside the corral. If a second man is present (otherwise the trouble is much greater), he holds the animal's head whilst the first puts on the horse-cloths and saddle, and girths the whole together. During this operation, the horse, from dread and astonishment at being thus bound round the waist, throws himself over and over again on the ground, and till beaten is unwilling to rise. At last, when the saddling is finished, the poor animal can hardly breathe from fear, and is white with foam and sweat. The man now prepares to mount by pressing heavily on the stirrup, so that the horse may not lose its balance; and at the moment he throws his leg over the animal's back, he pulls the slip-knot and the beast is free. The horse, wild with dread, gives a few most violent bounds, and then starts off at full gallop. When quite exhausted, the man by patience brings him back to the corral, where, reeking hot and scarcely alive, the poor beast is let free. Those animals which will not gallop away, but obstinately throw themselves on the ground, are by far the most troublesome.

In Chili a horse is not considered perfectly broken till he can be brought up standing in the midst of his full speed on any particular spot; for instance, on a cloak thrown on the ground; or again, will charge a wall and rearing, scrape the surface with his hoofs. I have seen an animal bounding with spirit, yet merely reined by a forefinger and thumb, taken at full gallop across a court-yard, and then made to wheel round the post of a verandah with great speed, but at so equal a distance that the rider with outstretched arm all the while kept one finger rubbing the post, then making a *demi-volte* in the air, with the other arm outstretched in a like manner, he wheeled round with astonishing force in the opposite direction. Such a horse is well broken; and although this at first may appear useless, it is far otherwise: it is only carrying that which is daily necessary into perfection. When a bullock is checked and caught by the lasso, it will sometimes gallop round and round in a circle, and the horse being alarmed at the great strain, if not well broken, will not readily turn like the pivot of a wheel. In consequence many men have been killed; for if a lasso once takes a twist round a man's body, it will instantly, from the power of the two animals, almost cut him in twain. On the same principal the races are managed. The course is only two or three hundred yards long, the desideratum being to have horses that can make a rapid dash. The race-horses are trained not only to stand with their hoofs touching a line, but to draw all four feet together, so as at the first spring to bring into play the full action of the hind quarters. In Chili I was told an anecdote, which I believe was true, and it offers a good illustration of the use of a well-broken animal. A respectable man riding one day met two others, one of whom was mounted on a horse, which he knew to have been stolen from himself. He challenged them; they answered by drawing their sabres and giving chase. The man on his good and fleet beast kept just ahead; as he passed a thick bush he wheeled round it, and brought up his horse to a dead check. The pursuers were obliged to shoot on one side and ahead. Then instantly dashing on right behind them, he buried his knife in the back of one, wounded the other, recovered his horse from the dying robber, and rode home!" Animals are so abundant in these countries that humanity is scarcely known. Mr. Darwin was one day riding in the Pampas with a very respectable "Estanciero," when his horse, being tired, lagged behind. The man often shouted to him to spur him, when I remonstrated that it was a pity, for the horse was quite exhausted, he cried: "Why not?—never mind. Spur him—it is *my* horse!" When after some difficulty he was made to understand that it was for the horse's sake that the spurs were not used, he exclaimed with great surprise: "Ah! Don Carlos, *qué cosa!*" The idea had never before entered his head.

In this country the powers of horses in swimming are but little tested, but in South America the case is different, as shown by an incident mentioned by Mr. Darwin. "I crossed the Lucia near its mouth, and was surprised to observe how easily our horses, although not used to swim, passed over a width of at least six hundred yards. On mentioning this at Monte Video, I was told that a vessel containing some mountebanks and their horses being wrecked in the Plata, one horse swam seven miles to the shore. In the course of the day I was amused by the dexterity with which a Guacho forced a restive horse to swim a river. He stripped off his clothes and jumping on its back, rode into the water till it was out of its depth; then slipping off over the crupper, he caught hold of the tail, and as often as the horse turned round, the man frightened it back by splashing water in its face. As soon as the horse touched the bottom on the other side the man pulled himself on, and was firmly seated, bridle in hand, before the horse gained the bank. A naked man on a naked horse is a fine spectacle. I had no idea how well the two animals suited each other. The tail of a horse is a very useful appendage. I have passed a river in a boat with four people in it, which was ferried across in the same way as the Guacho. If a man and horse have to cross a broad river, the best plan is for the man to catch hold of the pommel or mane, and help himself with the other arm."

The Turkuman horses are most highly prized in Persia, and are regularly trained by the Turkumans preparatory to their plundering expeditions. Before proceeding on a foray, these wild people knead a number of small hard balls of barley-meal, which, when wanted, they soak in water, and which serves as food both for themselves and their horses. It is a frequent practice with them in crossing deserts where no water is to be found, to open a vein in the shoulder of the horse and drink a little of his blood, which, according to their own opinion, benefits rather than injures the animal. It is confidently stated, that when in condition their horses have gone one hundred and forty miles within twenty-four hours; and it has been proved that parties of them were in the habit of marching from seventy to one hundred and five miles for twelve or fifteen days together without a halt. During Sir John Malcolm's first mission to Persia, he, when riding one day near a small encampment of Afshar families, expressed doubts to his Mehmander, a Persian nobleman, as to the reputed boldness and skill in horsemanship of their females. The Mehmander immediately called to a young woman of handsome appearance, and asked her in Turkish, if she was a soldier's daughter. She said she was. "And you expect to be a mother of soldiers?" She smiled. "Mount that horse," said he, pointing to one with a bridle, but without a saddle, "and show this European Elchee the difference between a girl of a tribe and a citizen's daughter." She instantly sprang upon the animal, and setting off at full speed, did not stop till she had reached the summit of a small hill in the vicinity, which was covered with loose stones. When there she waved her hand over her head, and came down the hill at the same rate at which she had ascended it. Nothing could be more dangerous than the ground over which she galloped; but she appeared quite fearless, and seemed delighted at having the opportunity of vindicating the females of her tribe from the reproach of being like the ladies of cities.

The *Shrubat-ur-Reech*, or *Drinkers of the Wind*, reared by the Mongrabins of the West, are shaped like greyhounds and as spare as a bag of bones, but their spirit and endurance of fatigue are prodigious. On one occasion the chief of a tribe was robbed of a favourite fleet animal of this race, and the camp went out in pursuit eight hours after the theft. At night, though the horse was not yet recovered, it was ascertained that the pursuers had headed his track, and would secure him before morning. The messenger who returned with this intelligence had ridden sixty miles in the withering heat of the desert without drawing bit. These animals are stated by Mr. Davidson, to be fed only once in three days, when they receive a large jar of camel's milk: this, with an occasional handful of dates, is their only food.

The fullest and most interesting account of the Arab horse has been written by General Daumas, and its value is greatly enhanced by containing a letter on the subject, written entirely by the celebrated Abd-el-Kadir, and a very remarkable document it is. According to this high authority, a perfectly sound Arab horse can, without difficulty, travel nearly thirty miles daily for three or four months, without resting a single day; and such a horse can accomplish fifty *parasangs* —not less than two hundred miles—in one day. When Abd-el-Kadir was with his tribe at Melonia, they made *razzias* in the Djebel-amour, pushing their horses at a gallop for five or six hours without drawing bridle, and they accomplished their expeditions in from twenty to twenty-five days. During all this time their horses ate only the corn carried by their riders, amounting to about eight ordinary meals. They often drank nothing for one or two days, and on one occasion were three days without water. The Arabic language is very epigrammatic, and the Arabs assign the reasons for instructing their horses early in these proverbs: "The lessons of infancy are graven in stone; but those of age disappear like the nests of birds." "The young branch without difficulty straightens itself— the large tree, never!" Accordingly, the instruction of the horse begins in the first year. "If," says the Emir, "the horse is not mounted before the third year, at the best he will only be good for the course; but *that* he has no need of learning—it is his natural faculty." The Arabs thus express the idea, "*Le djouad suivant sa race.*" The high-bred horse has no need of learning to run! The esteem of the Arab for his horse is conveyed in the following sentiment of the sage and saint, Ben-el-Abbas, which has been handed down from generation to generation: "Love thy horses— take care of them—spare thyself no trouble; by them comes honour, and by them is obtained beauty. If horses are abandoned by others, I take them into my family; my children share with them their bread; my wives cover them with their veils, and wrap themselves in their housings; I daily take them to the field of adventure; and, carried away by their impetuous course, I can fight with the most valiant."

General Daumas thus describes a combat between two tribes, drawn from life, for he enjoyed many opportunities of witnessing such scenes:—" The horsemen of the two tribes are in front, the women in the rear, ready to excite the combatants by their cries and applause: they are protected by the infantry, who also form the reserve. The battle is commenced by little bands of ten or fifteen horsemen, who hover on the flanks and seek to turn the enemy. The chiefs, at the head of a compact body, form the centre.

"Presently the scene becomes warm and animated —the young cavaliers, the bravest and best mounted, dash forward to the front, carried away by their ardour and thirst for blood. They uncover their heads, sing their war-songs, and excite to the fight by these cries, 'Where are those who have mistresses? It is under their eyes that the warriors fight to-day. Where are those who by their chiefs always boast of their valour? Now let their tongues speak loud, and not in those babblings. Where are those who run after reputation? Forward! forward! children of powder! Behold these sons of Jews—our sabres shall drink their blood— their goods we will give to our wives!' These cries inflame the horsemen—they make their steeds bound, and unsling their guns—every face demands blood— they mingle in the fray, and sabre-cuts are everywhere exchanged.

"However, one party has the worst of it, and begins to fall back on the camels which carry the women. Then are heard on both sides the women—on the one animating the conquerors by their cries of joy—on the other, seeking to stimulate the failing courage of their husbands and brothers by their screams of anger and imprecation. Under these reproaches the ardour of the vanquished returns, and they make a vigorous effort. Supported by the fire of the infantry who are in reserve, they recover their ground, and throw back their enemy into the midst of the women, who in their turn curse those whom just before they had applauded. The battle returns to the ground which lies between the females of the tribes. At last the party who have suffered most in men and horses, who have sustained the greatest loss, and have seen their bravest chiefs fall, take flight, in spite of the exhortations and prayers of those bold men who, trying to rally them, fly right and left, and try to recover the victory. Some warriors still hold their ground, but the general rout sweeps them off. They are soon by their women; then each, seeing that all is lost, occupies himself in saving that which is dearest; they gain as much ground as possible in their flight, turning from time to time to face the pursuing enemy. The conquerors might ruin them completely if the intoxication of their triumph did not build a bridge of gold for the vanquished, but the thirst of pillage disbands them. One despoils a foot-soldier—another a horseman. This one seizes a horse—that a negro. Thanks to this disorder the bravest of the tribe save their wives, and frequently their tents."

Before 1800 no political mission from a European

nation had visited the court of Persia for a century; but the English had fame as soldiers from the report of their deeds in India. An officer of one of the frigates which conveyed Sir John Malcolm's mission, who had gone ashore at Abusheher, and was there mounted on a spirited horse, afforded no small entertainment to the Persians by his bad horsemanship. The next day the man who supplied the ship with vegetables, and who spoke a little English, met him on board, and said, "Don't be ashamed, sir, nobody knows you: bad rider! I tell them you, like all English, ride well, but that time they see you, *you very drunk*." The worthy Persian thought it would have been a reproach for a man of a warlike nation not to ride well, but none for a European to get drunk.

Sir John Malcolm had taken with him to Persia a few couples of English foxhounds, intending them as a present to the heir-apparent, Abbas Mirza. Several excellent runs took place, greatly to the astonishment of the natives. One morning a fox was killed after a very hard chase; and whilst the rest of the party were exulting in their success, adding some two feet to a wall their horses had cleared, and relating wonderful hair-breadth escapes, Sir John was entertained by listening to an Arab peasant, who with animated gestures was narrating to a group of his countrymen all that he had seen of this noble hunt. "There went the fox," said he, pointing with a crooked stick to a clump of date-trees; "there he went at a great rate: I hallooed and halloed, but nobody heard me, and I thought he must get away; but when he was quite out of sight up came a large spotted dog, and then another and another; they all had their noses on the ground, and gave tongue, 'whow, whow, whow,' so loud that I was frightened. Away went these devils, who soon found the poor animal; after them galloped the Feringees, shouting and trying to make a noise louder than the dogs—no wonder they killed the fox among them; but it certainly is fine sport!"

Innumerable are the tales illustrative of the love of Arabs for their horses; but another anecdote mentioned by Sir John Malcolm places this in an amusing light. An English surgeon had been setting the broken leg of an Arab, who complained more of the accident which had befallen him than was thought becoming in one of his tribe: this the surgeon remarked to him, and his answer was truly characteristic,—"Do not think, Doctor, I should have uttered one word of complaint if my own high-bred colt, in a playful kick, had broke both my legs; but to have a bone broken by a brute of a jackass is too bad, and I *will* complain."

A touching incident is mentioned by Mungo Park

as having occurred whilst he, friendless and forlorn, was pursuing his weary journeyings far in the interior of Africa. The simple narrative tells its own tale of accumulated misery:—"July 29. Early in the morning my landlord observing that I was sickly, hurried me away, sending a servant with me as a guide to Kea. But though I was little able to walk, my horse was still less able to carry me, and about six miles to the east of Modiboo, in crossing some rough clayey ground he fell; and the united strength of the guide and myself could not place him again upon his legs. I sat down for some time beside this worn-out associate of my adventures; but finding him still unable to rise, I took off the saddle and bridle, and placed a quantity of grass before him. I surveyed the poor animal as he lay panting on the ground, with sympathetic emotion, for I could not suppress the sad apprehension that I should myself in a short time lie down and perish in the same manner of fatigue and hunger. With this foreboding I left my poor horse, and with great reluctance I followed my guide on foot along the bank of the river until about noon, when we reached Kea, which I found to be nothing more than a small fishing village."

Torn with doubt and perplexity, heavy of heart and weary in body, the unhappy traveller returned westward to Modiboo, after two days' journeying in company with a negro carrying his horse accoutrements. "Thus conversing," says he, "we travelled in the most friendly manner, until unfortunately we perceived the footsteps of a lion quite fresh in the mud near the river side. My companion now proceeded with great circumspection, and at last, coming to some thick underwood, he insisted that I should walk before him. I endeavoured to excuse myself by alleging that I did not know the road, but he obstinately persisted; and after a few high words and menacing looks, threw down the saddle and went away. This very much disconcerted me, for as I had given up all hopes of obtaining a horse, I could not think of encumbering myself with a saddle; and taking off the stirrups and girths, I threw the saddle into the river. The Negro no sooner saw me throw the saddle into the water, than he came running from among the bushes where he had concealed himself, jumped into the river, and by the help of his spear brought out the saddle, and ran away with it. I continued my course along the bank, but as the wood was remarkably thick, and I had reason to believe that a lion was at no great distance, I became much alarmed, and took a long circuit through the bushes to avoid him. About four in the afternoon I reached Modiboo, where I found my saddle; the guide, who had got there before me, being afraid that I should inform the king of his conduct, had brought the saddle with him in a canoe. While I was conversing with the dooty, and remonstrating with the guide for having left me in such a situation, I heard a horse neigh in one of the huts, and the dooty inquired with a smile if I knew who was speaking to me. He explained himself by telling me that my horse was still alive, and somewhat recovered from his fatigue." The happiness with which Park met his lost faithful steed may be conceived, for in him he had one friend left in the world.

Another lamented victim to African travel thus touchingly laments a grievous misfortune which befell him. Returning from an excursion to Konka, Major Denham writes:—" I was not at all prepared for the news which was to reach me on returning to our enclosure. The horse that had carried me from Tripoli to Moursuk and back again, and on which I had ridden the whole journey from Tripoli to Bornou, had died a very few hours after my departure for the lake. There are situations in a man's life in which losses of this nature are felt most keenly, and this was one of them. It was not grief, but it was something very nearly approaching to it; and though I felt ashamed of the degree of derangement which I suffered from it, yet it was several days before I could get over the loss. Let it, however, be remembered, that the poor animal had been my support and comfort—may I not say, companion?—through many a dreary day and night,—had endured both hunger and thirst in my service with the utmost patience,— was so docile, though an Arab, that he would stand still for hours in the desert while I slept between his legs, his body affording me the only shelter that could be obtained from the powerful influence of a noonday sun: he was the fleetest of the fleet, and ever foremost in the race."*

Captain Brown, in his 'Biographical Sketches of Horses,' gives the following interesting account of a circumstance that occurred at the Cape of Good Hope:—In one of the violent storms that often occur there, a vessel was forced on the rocks, and beaten to pieces. The greater part of the crew perished miserably, as no boat could venture to their assistance. Meanwhile a planter came from his farm to see the wreck, and knowing the spirit of his horse and his excellence as a swimmer, he determined to make a desperate effort for their deliverance, and pushed into the thundering breakers. At first both disappeared, but were soon seen on the surface. Nearing the wreck, he caused two of the poor seamen to cling to his boots, and so brought them safe to shore. Seven times did he repeat this perilous feat, and saved fourteen lives; but, alas! the eighth time,

* Narrative of Travels in Africa, by Major Denham.

the horse being much fatigued, and meeting with a formidable wave, the gallant fellow lost his balance, and was overwhelmed in a moment. He was seen no more, but the noble horse reached the land in safety.

Lieutenant Wellstead relates an adventure in his travels in Arabia, which illustrates the importance of being well mounted in that wild land:—" On my return from Obri to Suweik, contrary to the wish of the Bedouins, who had received intelligence that the Wahhábis were lurking around, I left the village where we had halted, alone, with my gun, in search of game. Scarcely had I rode three miles from the walls, when suddenly turning an angle of the rocks, I found myself within a few yards of a group of about a dozen horsemen, who lay on the ground, basking listlessly in the sun. To turn my horse's head and away was the work scarcely of an instant; but hardly had I done so when the whole party were also in their saddles in full cry after me. Several balls whizzed past my head, which Sayyid acknowledged by bounding forward like an antelope; he was accustomed to these matters, and their desire to possess him unharmed, alone prevented my pursuers from bringing him down. As we approached the little town, I looked behind me: a Sheikh better mounted than his followers was in advance, his dress and long hair streaming behind him, while he poised his long spear on high, apparently in doubt whether he was sufficiently within range to pierce me. My good stars decided that he was not; for reining up his horse he rejoined his party, whilst I gained the walls in safety! The day before Sayyid came into my hands he had been presented to the Imaum by a Nejd sheikh; reared in domesticity, and accustomed to share the tent of some Arab family, he possessed, in an extraordinary degree, all the gentleness and docility, as well as the fleetness, which distinguish the pure breed of Arabia. To avoid the intense heat and rest their camels, the Bedouins frequently halted during my journey for an hour about mid-day. On these occasions Sayyid would remain perfectly still while I reposed on the sand, screened by the shadow of his body. My noon repast of dates he always looked for and shared. Whenever we halted, after unsaddling him and taking off his bridle with my own hands, he was permitted to roam about the encampment without control. At sunset he came for his corn at the sound of my voice, and during the night, without being fastened, he generally took up his quarters at a few yards from his master. During my coasting voyages along the shore, he always accompanied me, and even in a crazy open boat from Muskat to India. My health having compelled me to return to England overland, I could not in consequence bring Sayyid with me. In parting with this attached and faithful creature, so long the companion of my perils and wanderings, I am not ashamed to acknowledge, that I felt an emotion similar to what is experienced in being separated from a tried and valued friend."

Among the North American Indians the Camanchees take the first rank as equestrians; racing, indeed, is with them a constant and almost incessant exercise, and a fruitful source of gambling. Among their feats of riding is one described by Mr. Catlin as having astonished him more than anything in the way of horsemanship he had ever beheld; and it is a stratagem of war familiar to every young man in the tribe. At the instant he is passing an enemy, he will drop his body upon the opposite side of the horse, supporting himself with his heel upon the horse's back. In this position, lying horizontally, he will hang whilst his horse is at its fullest speed, carrying with him his shield, bow and arrows, and lance fourteen feet long, all or either of which he will wield with the utmost facility, rising and throwing his arrows over the horse's back, or under his neck, throwing himself up to his proper position, or changing to the other side of the horse if necessary. The actual way in which this is done is as follows:—A short hair halter is passed under the neck of the horse, and both ends tightly braided into the mane, leaving a loop to hang under the neck and against the breast. Into this loop the rider drops his elbow suddenly and fearlessly, leaving his heel to hang over the back of the horse to steady him and enable him to regain the upright position.

The following very singular custom prevails among the tribe of North American Indians known as the *Foxes*. Of it Mr. Catlin was an eye-witness. "When," says he, "General Street and I arrived at Kee-o-kuk's village, we were just in time to see this amusing scene on the prairie, a little back of his village. The Foxes, who were making up a war-party to go against the Sioux, and had not suitable horses enough by twenty, had sent word to the 'Sacs' the day before, according to ancient custom, that they were coming on that day, at a certain hour, to 'smoke' that number of horses, and they must not fail to have them ready. On that day, and at the hour, the twenty young men who were beggars for horses were on the spot, and seated themselves on the ground in a circle, where they went to smoking. The villagers flocked round them in a dense crowd, and soon after there appeared on the prairie, at half a mile distance, an equal number of young men of the Sac tribe, who had agreed each to give a horse, and who were then galloping them round at full speed; and gradually, as they went around in a circuit, coming nearer to the centre, until they were at last close around the ring of young fellows seated on the ground. Whilst dashing about thus, each one with a heavy whip in his hand, as he came within reach of the group on the ground, selected the one to whom he decided to present his horse, and as he passed gave him the most tremendous cut with his lash over the naked shoulders; and as he darted around again, he plied the whip as before, and again and again with a violent 'crack,' until the blood could be seen trickling down over his naked shoulders, upon which he instantly dismounted, and placed the bridle and whip in his hands, saying, 'Here, you are a beggar; I present a horse, but you will carry my mark on your back.' In this manner they were all in a little while 'whipped up,' and each had a good horse to ride home and into battle. His necessity was such that he could afford to take the stripes and the scars as the price of the horse, and the giver could afford to make the present for the satisfaction of putting his mark on the other, and of boasting of his liberality."

Mr. Catlin gives an interesting account of his faithful horse "Charley," a noble animal of the Camanchee wild breed, which had formed as strong an attachment for his master as his master for him. The two halted generally on the bank of some little stream, and the first thing done was to undress

Charley, and drive down the picket to which he was fastened, permitting him to graze over a circle limited by his lasso. On a certain evening, when he was grazing as usual, he managed to slip the lasso over his head, and took his supper at his pleasure as he was strolling round. When night approached, Mr. Catlin took the lasso in hand, and endeavoured to catch him, but he continually evaded the lasso until dark, when his master abandoned the pursuit, making up his mind that he should inevitably lose him, and be obliged to perform the rest of the journey on foot. Returning to his bivouac, in no pleasant state of mind, he lay down on his bearskin and went to sleep. In the middle of the night he awoke whilst lying on his back, and, half opening his eyes, was petrified at beholding, as he thought, the huge figure of an Indian standing over him, and in the very act of stooping to take his scalp! The chill of horror that paralyzed him for the first moment held him still till he saw there was no need of moving; that his faithful horse had played shy till he had filled his belly, and had then moved up from feelings of pure affection, and taken his position with his fore feet at the edge of his master's bed, and his head hanging over him, in which attitude he stood fast asleep.

When sunrise came the traveller awoke and beheld his faithful servant at a considerable distance, picking up his breakfast among the cane brake at the edge of the creek. Mr. Catlin went busily to work to prepare his own, and having eaten it, had another half hour of fruitless endeavours to catch Charley, who in the most tantalizing manner would turn round and round, just out of his master's reach. Mr. Catlin, recollecting the evidence of his attachment and dependence afforded by the previous night, determined on another course of proceeding, so packed up his traps, slung the saddle on his back, trailed his gun, and started unconcernedly on his route. After advancing about a quarter of a mile, he looked back, and saw Master Charley standing with his head and tail very high, looking alternately at him and at the spot where he had been encamped, and had left a little fire burning. Thus he stood for some time, but at length walked with a hurried step to the spot, and seeing everything gone, began to neigh very violently, and, at last, started off at fullest speed and overtook his master, passing within a few paces of him, and wheeling about at a few rods distance, trembling like an aspen-leaf. Mr. Catlin called him by his familiar name, and walked up with the bridle on his hand, which was put over Charley's head as he held it down for it, and the saddle was placed on his back as he actually stooped to receive it; when all was arranged, and his master on his back, off started the noble animal as happy and contented as possible.

Many of the American prairies swarm not only with buffaloes, but with numerous bands of wild horses, proud and playful animals, rejoicing in all the exuberance of freedom, and sweeping the earth with their flowing manes and tails. The usual mode of taking them by the North American Indians is by means of the lasso. When starting for the capture of a wild horse, the Indian mounts the fleetest steed he can get, and coiling the lasso under his arm, starts off at full speed till he can enter the band, when he soon throws the lasso over the neck of one of the number. He then instantly dismounts, leaving his own horse, and runs as fast as he can, letting the lasso pass out gradually and carefully through his hands, until the horse falls half-suffocated, and lies helpless on the ground. The Indian now advances slowly towards the horse's head, keeping the lasso tight upon his neck until he has fastened a pair of hobbles upon his fore feet; he now loosens the lasso, and adroitly casts it in a noose round the lower jaw, the animal, meanwhile, rearing and plunging. Advancing warily hand over hand, the man at length places his hand over the animal's eyes and on its nose, and then breathes into its nostrils, on which the horse becomes so docile and thoroughly conquered, that his captor has little else to do but to remove the hobbles from his feet, and ride or lead it into camp.

A remarkable instance of the confidence of a horse in a firm rider, and his own courage, was conspicuously evinced in the case of an Arab, mentioned by Lieutenant-Colonel Hamilton Smith. General Sir Robert Gillespie happened, when mounted on this animal, to be present on the race-course of Calcutta during one of the great Hindoo festivals, when several hundred thousand people had assembled. On a sudden an alarm was given that a tiger had escaped from his keepers. Sir Robert immediately snatched a boar-spear, and rode to attack this formidable enemy. The tiger was probably confounded by the crowd, but the moment he perceived Sir Robert, he crouched to spring at him. At that instant, the gallant soldier on his gallant steed leapt right over the tiger, Sir Robert striking the spear through the animal's spine! This horse was a small grey, but he possessed another which has become almost historical. It was a favourite black charger, bred at the Cape of Good Hope, and carried with him to India. When the noble soldier fell at the storming of Kalunga, this charger was put up for sale, and after great competition was knocked down to the privates of the 8th Dragoons, who actually contributed their prize-money to the amount of £500 to retain this memorial of their beloved commander. The beautiful charger was always led at the head of the regiment on a march, and at the station at Cawnpore took his ancient post at the colour-stand, where the salute of passing squadrons was given at drill, and on reviews. When the regiment was ordered home, the funds of the privates running low, he was bought by a gentleman, who provided a paddock for him, where he might pass the remainder of his days in comfort; but when the corps had departed, and the sound of the trumpet was heard no more, the gallant steed pined, refused his food, and on the first opportunity, being led out for exercise, he broke from his groom, galloped to his ancient station on parade, neighed loudly again and again; and there, on the spot where he had so often proudly borne his beloved master, he dropped down and died!

Before the battle of Corunna, it being found impossible to embark the horses of the cavalry in the face of the enemy, they were ordered to be shot to prevent their being distributed among the French cavalry. The poor animals, the faithful companions of the troopers in many a weary march and hard-fought skirmish, stood trembling as they saw their companions fall one after the other, and by their piteous looks seemed to implore mercy, till the duty imposed upon the dragoons entrusted with the execution of the order became unbearable, and the men turned away from their task with scalding tears;

hence the French obtained a considerable number unhurt, and among them several belonging to officers, who rather than destroy their faithful chargers, had left them with billets attached, recommending them to the kindness of the enemy.

We will conclude with an anecdote related of a son of a late church dignitary, whose taste lay more in the sports of the field and the 'Stud Book,' than in Cudworth's 'Intellectual System of the Universe' or such light reading. He was on an important occasion to meet the Bishop of L—— at dinner, and as it was desirable that a favourable impression should be made upon his lordship, his father begged he would be agreeable to the bishop, and do his best to draw him out, as he was strong in Biblical lore. Matters went on pleasantly enough during the early part of the banquet, our friend saying little, but watching his opportunity for a charge. At length a pause took place, and he thus addressed the bishop, the company listening. "Might I venture to ask your Lordship a question relative to a point mentioned in the Old Testament which has puzzled me a good deal?" "Oh, certainly—most happy!" said the dignitary, feeling quite in his element. "Then I should be glad to have your Lordship's opinion as to how long it took Nebuchadnezzar *to get into condition after he had been out to grass?*"

The bishop was *not* in his element.

CHAPTER VI.

NEW SPECIES.—ANCIENT KNOWLEDGE OF THE GIRAFFE.—PLINY.—
LORENZO THE MAGNIFICENT.—GIRAFFE AT PARIS.—GEORGE IV.
—CAPTURE OF GIRAFFES.—A SINGULAR PROCESSION.—GREAT
ATTRACTION.—THE FIRST BIRTH.—ACCIDENT.—A DAINTY DISH.
—TONGUE OF GIRAFFE.—PETTY LARCENY.—THE PEACOCK UN-
TAILED.—MOVEMENTS OF GIRAFFE.—GENTLENESS.—SIR CORN-
WALLIS HARRIS. — FIRST VIEW. — A DISAPPOINTMENT.—VEX-
ATIOUS INCIDENTS. — COOLNESS OF HOTTENTOTS. — ANOTHER
TROOP.—HOSTILE RHINOCEROS.—HOT CHASE.—THE DEATH.—A
DESERTER.—ADAPTATION OF FORM.—REFLECTIONS.

"The admirablest and fairest beast that ever I saw, was a *jarraff*, as tame as a domesticall deere, and of a reddish deere colour, white brested, and cloven footed, he was of a very great height, his fore legs longer than the hinder, a very long necke, and headed like a camell, except two stumpes of horn upon his head. This fairest animall was sent out of Ethiopia to this great Turke's father for a present; two Turks, the keepers of him, would make him kneele, but not before any Christian for any money. An elephant that stood where this fair beast was, the keepers would make to stand with all his four legges, his feet close together upon a round stone, and alike to us to bend his fore legges."—
JOHN SANDERSON, 1591.

OF the many features which will hereafter stamp the nineteenth century as "Centuria Mirabilissima," not the least will be the vast number of animals and birds introduced into Europe, and the great stride made in our knowledge of Natural History during its progress. The precise date of the extinction of a genus or a species has interest; the dodo of the Mauritius and the dinornis of New Zealand have disappeared within the historical period, and there is no reason to suppose that such gaps have been, or will be, filled up by new creations. Second only in interest to the occurrence of these blanks in the list of living inhabitants of the surface of this globe is the record of the introduction of a new race into a part of our planet where it was previously unknown. In such instances the last twenty years have been prolific; the graceful bower-birds and the *Tallegalla* or mound-raising birds, those wondrous denizens of the Australian wilderness, may now be seen in the Regent's Park, for the first time in this hemisphere. For the first time, also, the wart-hog of Africa there roots, and the hippopotamus displays his quaint gambols; and that "fairest animall," the giraffe, is now beheld in health and vigour, a naturalized inhabitant of Great Britain.

In the modern versions of the Old Testament, the fifth verse of the fourteenth chapter of Deuteronomy, which enumerates animals permitted to be eaten by the Israelites, mentions "the hart and the roebuck, and the fallow deer, and the wild goat, and the pygarg, and the wild ox, and the chamois." It has been objected by able commentators, that instead of the chamois the giraffe was implied; and the commentator in the Pictorial Bible says, "The Arabic version understood that the giraffe was meant here, which is very likely to have been the case; for the chamois is not met so far to the southward as Egypt and Palestine. The giraffe, or cameleopard, is a singular as well as beautiful creature, found in the central parts of Africa. The Jews had probably many opportunities of becoming acquainted with the animal while in Egypt, as had also the Seventy, who resided there, and who indicate it in their translation of the Hebrew name. Belzoni also notices the giraffe on the wall of the sekos of the Memnonium, and on the back of the temple of Ermeuts. Thinking it possible that a representation of this animal might be found either in the triumphal processions or hunting scenes of the Ancient Assyrians, we have examined the Nineveh marbles in the British Museum, but although elephants and very many animals foreign to Assyria are delineated, the giraffe is not among them.

Sylla held the office of quæstor in Numidia, and the Prænestine pavement, generally referred to him, represents the giraffe both grazing and browsing; but the living animal does not seem to have been brought to Rome before the time of Julius Cæsar: such at least states Pliny, who may be regarded as a good authority on that matter. "Two other kinds of beasts there be that resemble in some sort the camels; the one is called of the Æthiopians, the nabis, necked like a horse, for leg and hoofe not unlike the bœufe, headed directly like a camell, beset with white spots upon a red ground, whereupon it taketh the name of camelopardalus; and the first time that it was seen at Rome was in the games Circenses, set out by Cæsar Dictator. Since which time he comes now and then to Rome to be looked upon more for sight than for any wild nature that hee hath; whereupon some call her the savage sheepe."* It is pretty evident from this that Pliny himself never had the good fortune to see one of these most miscalled "savage sheepe;" but Gordian III. had ten living giraffes at one time, and though from their gentleness ill-suited for the circus, they were well adapted to form striking features in the triumphal processions, for which purpose they were doubtless used.

Towards the end of the fifteenth century flourished Lorenzo de' Medici the Magnificent: this illustrious man, like numerous other eminent characters, sought relaxation from the cares of state concerns, at a country-seat, and his villa of Poggio Cajano was a favoured residence. The lakes of this charming domain abounded with choice water-fowl, the woods

* Holland's Plinie's Naturall Historie, B. VIII. c. xviii.

with pheasants and peacocks from Sicily, and a collection of rare exotic animals increased the interest of the spot. Lorenzo was passionately fond of them,* and the Soldan of Egypt, hearing of his zoological taste, sent him a present of a giraffe; this animal throve greatly, and is said to have become very familiar with the inhabitants of Florence, stretching up its long neck to the first-floors of the houses to implore a meal of apples, of which it was passionately fond. The portrait of this giraffe is, we believe, still extant in the frescoes which adorn the villa of Poggio Cajano.

Three centuries and a half elapsed before another specimen of this animal was seen in Europe, and naturalists began to be sceptical as to its existence, when the liberality of another Pasha of Egypt dispelled all doubts. The governor of Sennaar having sent two young giraffes to the Pasha, his highness graciously determined to present them to the sove-

* Valori states that, "among the numerous horses kept by Lorenzo was a beautiful roan that on every occasion bore away the prize; when his horse happened to be sick or wearied with the course, he refused all nourishment except from the hands of Lorenzo, at whose approach he testified his pleasure by neighing and movements of welcome, even when lying on the ground; so that is not to be wondered at," says this author, by a kind of commendation rather more striking than just, "that Lorenzo should be the delight of mankind, when even the brute creation expressed an affection for him."

reigns of England and of France, and desired their consuls to draw lots for the choice. The result was, that the most vigorous went to Paris, and the feeblest came to England. Great were the rejoicings in the French capital at the arrival of this stranger; all classes were in a commotion at the event; the animal was conveyed in a sort of triumphal procession to the Jardin des Plantes: the picture-shops teemed with her portraits, and for several months little else was talked of. Every fashion was *à la giraffe;*—ladies appeared, not

> "Spotted like the pard,"

but like the cameleopard; gentlemen bore her portrait on their handkerchiefs, whilst the vests of the dandies, and even their pantaloons, bore fanciful combinations of her spots. Ati, her keeper, a tall, fine-looking Arab, partook of her popularity; at first he appeared in the streets attired in the turban, vest, and full trousers of the East, but was so tormented by the attentions of the *gamins*, who were continually assailing him with "*Ati! Ati! comment va la giraffe?*" that, except when on duty with his charge, he was fain to affect the short-waisted coats and bell-crowned hats then in vogue. Every Sunday evening this worthy gentleman was to be seen at one of the *guinguettes* in the neighbourhood of the Jardin des Plantes, dancing with all his might, and playing the agreeable to the *grisettes*, with whom he was extremely popular.

The giraffe destined for our sovereign was conveyed to Malta under the charge of two Arabs, and was from thence forwarded to London in the 'Penelope,' which arrived on the 11th of August, 1827. She was conveyed to Windsor two days afterwards, and was kept in the royal menagerie at the Sandpit Gate. George IV. took much interest in this animal, visiting her generally twice or thrice a week, and sometimes twice a day. It would have been better if he had left her to the management of the keepers; but acting on some vague instructions left by the Arabs, his Majesty commanded that she should be fed on milk alone, a most unnatural diet when the animal had attained the age of two years. From this cause, and in consequence of an injury which she had received during her journey from Sennaar to Cairo, the giraffe became so weak as to be unable to stand: a lofty triangle was built, and the animal kept suspended on slings to relieve its limbs from the support of its weight. The apparatus was provided with wheels; and, in order that she might have exercise, it was pushed along by men, her feet just moving and touching the ground. It may well be supposed that such an artificial existence could not be prolonged to any great length of time; and although the giraffe lived between two and three years, and grew eighteen inches in height, she gradually sank and died in the autumn of 1829, to the great regret of the king. Her body was dissected by the sergeant-surgeon, Sir Everard Home, and an account thereof published by him.

When these two giraffes were at Alexandria, previous to their embarkation, they were one day ordered by M. Acerbi to be led up and down the square in front of his house; among the crowd collected to enjoy this novel spectacle were some Bedouins of the Desert. One of them was asked whether he had ever seen similar animals before? He replied he had not: and M. Acerbi asked him in Arabic, " Do they please you?" to which he replied, " Mustaib"—I do not like them. On being pressed for the reasons of his disapproval, he answered, "that it did not carry like a horse, it did not serve for field labours like an ox, did not yield hair like a camel, nor flesh and milk like a goat, and on this account it was not to his liking."

Those who frequented the British Museum in the days of Montague House, shortly before the present building was erected, will remember a hairless stuffed giraffe, which stood at the top of the stairs, mounting sentry, as it were, over the principal door. This miserable skin was interesting, as being the remains of the first entire specimen recorded. Its history was as follows. The late Lady Strathmore sent to the

Cape to collect rare flowers and trees, a botanist of the name of Paterson, who seems to have penetrated a considerable distance into the interior—sufficiently far at least to have seen a group of six giraffes. He was so fortunate as to kill one, and brought the skin home for Lady Strathmore; her ladyship presented it to the celebrated John Hunter, and it formed part of the Hunterian collection until a re-arrangement of that museum took place on its removal to the present noble hall in the College of Surgeons. This stuffed specimen, with many others of a similar description, was handed over to the British Museum, and for some years occupied the situation on the landing above mentioned; being regarded as "rubbish," it was destroyed, and the "stuffing" used to expand some other skin. There are now, however, two noble stuffed specimens in the first zoological room of the Museum; one, especially remarkable for its dark-brown spots, is no less than eighteen feet in height. It is from the southern parts of Africa, and was presented by the late Earl of Derby; the other was one of the giraffes brought by M. Thibaut to the Zoological Gardens.

The Zoological Society having made known its wish to possess living specimens of the giraffe, the task of procuring them was undertaken by M. Thibaut, who having had twelve years' experience in African travel, was well qualified for the arduous pursuit.

M. Thibaut quitted Cairo in April, 1834, and after sailing up the Nile as far as Wadi Halfa, the second cataract, took camels and proceeded to Debbat, a province of Dongolah, whence he started for the Desert of Kordofan. Being perfectly acquainted with the locality, and on friendly terms with the Arabs, he attached them still more by the desire of profit; all were desirous of accompanying him in pursuit of the giraffes, for, up to that time, they had hunted them solely for the sake of the flesh, which they ate, and the skin, of which they made bucklers and sandals. The party proceeded to the south-west of Kordofan, and in August were rewarded by the sight of two beautiful giraffes; a rapid chase of three hours, on horses accustomed to the fatigues of the desert, put them in possession of the largest of these noble animals; unable to take her alive, the Arabs killed her with blows of the sabre, and cutting her to pieces, carried the meat to their head-quarters, which had been established in a wooded situation, an arrangement necessary for their own comfort, and to secure pasturage for their camels. They deferred till the following day the pursuit of the motherless young one, knowing they would have no difficulty in again discovering it. The Arabs quickly covered the live embers with slices of the meat, which M. Thibaut pronounces to be excellent.

On the following morning the party started at daybreak in search of the young giraffe, of which they had lost sight not far from the camp. The sandy desert is well adapted to afford indications to a hunter, and in a very short time they were on the track of the object of their pursuit: they followed the traces with rapidity and in silence, lest the creature should be alarmed whilst yet at a distance; but after a laborious chase of several hours through brambles and thorny trees, they at last succeeded in capturing the coveted prize.

It was now necessary to rest for three or four days, in order to render the giraffe sufficiently tame, during which period an Arab constantly held it at the end of a long cord; by degrees it became accustomed to the presence of man, and was induced to take nourishment, but it was found necessary to insert a finger into its mouth to deceive it into the idea that it was with its dam: it then sucked freely. When captured, its age was about nineteen months. Five giraffes were taken by the party, but the cold weather of December 1834 killed four of them in the desert, on the route to Dongolah; happily that first taken survived, and reached Dongolah in January 1835, after a sojourn of twenty-two days in the desert. Unwilling to leave with a solitary specimen, M. Thibaut returned to the desert, where he remained three months, crossing it in all directions, and frequently exposed to great hardships and privations; but he was eventually rewarded by obtaining three giraffes, all smaller than the first. A great trial awaited them, as they had to proceed by water the whole distance from Wadi Halfa to Cairo, and thence to Alexandria and Malta, besides the voyage to England. They suffered considerably at sea during a passage of twenty-four days in very tempestuous weather, and on reaching Malta in November, were detained in quarantine twenty-five days more; but despite of all these difficulties, they reached England in safety, and on the 25th of May were conducted to the Gardens. At daybreak the keepers and several gentlemen of scientific distinction, arrived at the Brunswick Wharf, and the animals were handed over to them. The distance to the Gardens was not less than six miles, and some curiosity, not unmingled with anxiety, was felt as to how this would be accomplished. Each giraffe was led between two keepers, by means of long reins attached to the head; the animals walked along at a rapid pace, generally in advance of their conductors, but were perfectly tractable. It being so early in the morning, few persons were about, but the astonishment of those who did behold the unlooked-for procession, was ludicrous in the extreme. As the giraffes stalked by, followed by M. Thibaut and others, in Eastern costume, the worthy policemen and early coffee-sellers stared with amazement, and a few revellers, whose reeling steps proclaimed their dissipation, evidently doubted whether the strange figures they beheld were real flesh and bone, or fictions conjured up by their potations; their gaze of stupid wonder indicating that, of the two, they inclined to the latter opinion. When the giraffes entered the park, and first caught sight of the green trees, they became excited, and hauled upon the reins, waving the head and neck from side to side, with an occasional caracole and kick-out of the hind legs, but M. Thibaut contrived to coax them along with pieces of sugar, of which they were very fond, and he had the satisfaction of depositing his valuable charges, without accident or misadventure, in the sanded paddock prepared for their reception.

The sum agreed on with M. Thibaut was £250 for the first giraffe he obtained, £200 for the second, £150 for the third, and £100 for the fourth, in all £700; but the actual cost to the Society amounted to no less than £2386. 3s. 1d., in consequence of the heavy expenses of freight, conveyance, etc.

During the following months of June and July, the giraffes excited so much interest, that as much as £120 was sometimes taken at the Gardens in one day, and the receipts reached £600 in the week; they then decreased, and never, until the arrival of the hippopotamus, attained anything like that sum

P

again. Shortly after their arrival, one of the animals struck his head with such force against the brickwork of the house, whilst rising from the ground, that he injured one of his horns, and probably his skull, as he did not long survive. Guiballah died in October, 1846, and Selim in January, 1849; Zaida, that worthy old matron, is, we believe, still alive, and may be recognized by her very light colour.

An unusual birthday *fête* was celebrated on the 9th of June, 1839, when Zaida presented the society with the first giraffe ever born in Europe; but, alas! it only survived nine days. A spirited water-colour sketch was made of the dam and young one, when a day old, by that able artist, the late Robert Hills; and we recently had an opportunity of seeing this interesting memento. Two years afterwards, a second was born, and throve vigorously; this fine animal was sent to the Zoological Gardens at Dublin, in 1844. It was rather a ticklish proceeding, but was managed as follows:—He was taken very early in the morning to Hungerford Market, where a lighter with tackles had been previously arranged. With some dexterity slings were placed under him, and to his great astonishment, he was quickly swung off his feet, and hoisted by a crane into the lighter, and from the lighter, by tackle, on board the deck of the steamer; he had a fine passage, and was welcomed with enthusiasm by the warm-hearted Hibernians, and is, we believe, still one of the chief ornaments of the Dublin Gardens. Another remarkably fine male, named *Abbas Pasha*, was born in February, 1849, and is thriving in great vigour in the Gardens at Antwerp.

The giraffes at present (1852) in the Regent's Park are *Zaida*, with her offspring *Alfred* and *Ibrahim Pasha*, *Alice*, presented by his highness Ibrahim Pasha, and *Jenny Lind* purchased by Mr. Murray. With the exception of *Ibrahim Pasha*, these are exceedingly good-tempered, but this fine animal is obliged to be kept separate, as he is very apt to fight with his brother. Their mode of fighting is peculiar; they stand side by side, and strike obliquely with their short horns, denuding the parts struck, to the magnitude of a hand. One of them met with an awkward accident some time ago, which, had it not been for the presence of mind of Mr. Hunt, the head keeper, who has especial charge of these animals, might have been attended with fatal consequences. In rising quickly from the ground, the giraffe struck the wall with such force that one of the horns was broken, and bent back flat upon the head; Hunt seeing this, tempted him with a favourite dainty with one hand, and taking the opportunity whilst his head was down, grasped the fractured horn, and pulled it forward into its natural position; union took place, and no ill effects followed. We may here remark, that the horns are distinct bones, united to the frontal and parietal bones by a suture, and exhibiting the same structure as other bones. The protuberance on the forehead is not a horn (as supposed by some), but merely a thickening of the bone. The horns of the male are nearly double the size of those of the female, and their expanded bases meet in the middle line of the skull, whereas, in the female, the bases are two inches apart.

Each of the giraffes eats daily eighteen pounds of clover hay, and the same quantity of a mixed vegetable diet, consisting of turnips, mangel-wurzel, carrots, barley, and split beans; in spring they have green tares and clover, and are exceedingly fond of onions. It was curious to see the impatience they manifested in our presence when a basket of onions was placed in view; their mouths watered to a ludicrous and very visible extent; they pawed with their fore legs, and rapidly paced backwards and forwards, stretching their long necks, and sniffing up the pungent aroma with eager satisfaction. Each drinks about four gallons of water a day.

Soon after the arrival of the giraffes at the Regent's Park, Mr. Warwick obtained three for Mr. Cross, of the Surrey Gardens. These were exhibited in an apartment in Regent Street, in the evening as well as by day; their heads almost touched the ceiling, and the room being lighted with gas, they were fully exposed to the influence of foul air, and, as might be expected, did not long survive. We have understood that Mr. Wombwell also purchased some giraffes on speculation, but was not more fortunate than Mr. Cross; indeed, of all animals, these are least adapted for the confinement and fetid atmosphere of a travelling menagerie.

It has been stated that giraffes utter no sound; we have, however, heard *Ibrahim Pasha* make a sort of grunt, or forcible expiration, indicating displeasure, and the little one which died bleated like a calf.

The extensibility, flexibility, and extraordinary command which the giraffe possesses over the movements of its tongue had long attracted notice, but it was reserved for Professor Owen to point out their true character. Sir Everard Home, who had examined the giraffe which died at Windsor, described the wonderful changes of size and length which occur in the tongue, as resulting from vascular action, the blood-vessels being at one time loaded, at another empty; but the Hunterian Professor proved that the movements of the tongue are entirely due to muscular action, and adds the following interesting remarks :—" I have observed all the movements of the tongue, which have been described by previous authors. The giraffe being endowed with an organ so exquisitely formed for prehension, instinctively puts it to use in a variety of ways, while in a state of confinement. The female in the Garden of Plants at Paris, for example, may frequently be observed to amuse itself by stretching upwards its neck and head, and, with the slender tongue, pulling out the straws which are plaited into the partition separating it from the contiguous compartment of its enclosure. In our own menagerie many a fair lady has been robbed of the artificial flower which adorned her bonnet, by the nimble, filching tongue of the object of her admiration. The giraffe seems, indeed, to be guided more by the eye than the nose in the selection of objects of food; and, if we may judge of the apparent satisfaction with which the mock leaves and flowers so obtained are masticated, the tongue would seem by no means to enjoy the sensitive in the same degree as the motive powers. The giraffes have a habit, in captivity at least, of plucking the hairs out of each other's manes and tails and swallowing them. I know not whether we must attribute to a fondness for epidermic productions, or to the tempting green colour of the parts, the following ludicrous circumstance, which happened to a fine peacock, which was kept in the giraffe's paddock. As the bird was spreading his tail in the sunbeams, and curvetting in presence of his mate, one of the giraffes stooped his long neck, and entwining his flexible tongue round a bunch of the gaudy plumes, suddenly lifted the bird into the air, then giving him a shake, disengaged five or six of the tail-feathers, when down fluttered the astonished peacock, and scuffled off with the remains of his train dragging humbly after him."*

The natural food of the giraffe is the leaves, tender shoots, and blossoms, of a singular species of mimosa, called by the colonists *kameel-doorn*, or giraffe-thorn, which is found chiefly on dry plains and sandy deserts. The great size of this tree, together with its thick and spreading top, shaped like an umbrella, distinguish it at once from all others. The wood, of a dark red colour, is exceedingly hard and weighty, and is extensively used by the Africans in the manufacture of spoons and other articles, many being ingeniously fashioned with their rude tools into the form of the giraffe.

The class to which the giraffe belongs is the deer tribe. It is in fact, as pointed out by Professor Owen, a modified deer; but the structure by which so large a ruminant is enabled to subsist in the tropical regions of Africa, by browsing on the tops of trees, disqualifies it for wielding antlers of sufficient strength and size to serve as weapons of offence. The annual shedding of the formidable antlers of the full-grown

* Transactions of the Zoological Society.

buck has reference to the preservation of the younger and feebler individuals of his own race; but, as the horns of the giraffe never acquire the requisite development to serve as weapons of attack, their temporary removal is not needed.

When looking at a giraffe, it is difficult to believe that the fore-legs are not longer than the hind-legs. They are not so, however, for the greater apparent length results from the remarkable depth of the chest, the great length of the processes of the anterior dorsal vertebræ, and the corresponding length and position of the shoulder-blade, which is relatively the longest and narrowest of all mammalia. In the simple walk the neck is stretched out in a line with the back, which gives them an awkward appearance; this is greatly diminished when the animals commence their undulating canter. In the canter the hind-legs are lifted alternately with the fore, and are carried outside of and beyond them, by a kind of swinging movement; when excited to a swifter pace, the hind legs are often kicked out, and the nostrils are then widely dilated. The remarkable gait is rendered still more automaton-like by the switching at regular intervals of the long black tail which is invariably curled above the back, and by the corresponding action of the neck, swinging as it does like a pendulum, and literally giving the creature the appearance of a piece of machinery in motion. The tail of the giraffe is terminated by a bunch of wavy hair, which attains a considerable length, but the longest hairs are those which form a fringe, extending about three inches on its under side. Two of these in our possession, from the tail of *Alfred*, are each rather more than four feet two inches in length. This long wisp of hair must be of great service in flicking off flies and other annoyances.

Major Gordon relates an anecdote of a giraffe slain by himself, which illustrates the gentle, confiding disposition of these graceful creatures. Having been brought to the ground by a musket-ball, it suffered the hunter to approach, without any appearance of resentment, or attempt at resistance. After surveying the crippled animal for some time, the Major stroked its forehead, when the eyes closed as if with pleasure, and it seemed grateful for the caress. When its throat was cut, preparatory to taking the skin, the giraffe, whilst struggling in the last agonies, struck the ground convulsively with its feet with immense force, as it looked reproachfully on its assailant with its fine eyes fast glazing with the film of death, but made no attempt to injure him.

Some of the best and most animating accounts of giraffe-hunts are contained in the works of Sir W. Cornwallis Harris and Mr. R. G. Cumming. Of that magnificent folio, 'Portraits of the Game and Wild Animals of South Africa,' by the former of these gallant sportsmen, we cannot speak too highly; it is equal in many respects to the truly superb folios of Mr. Gould. From it we extract the following spirit-stirring adventures.

"It was on the morning of our departure from the residence of his Amazoola Majesty, that I first actually saw the giraffe. Although I had been for weeks on the tiptoe of expectation, we had hitherto succeeded in finding the gigantic footsteps only of the tallest of all the quadrupeds upon the earth; but at dawn of that day, a large party of hungry savages, with four of the Hottentots on horseback, having accompanied us across the Mariqua in search of elands, which were reported to be numerous in the neighbourhood, we formed a long line, and having drawn a great extent of country blank, divided into two parties, Richardson keeping to the right, and myself to the left. Beginning at length to despair of success, I had shot a harte-beeste for the savages, when an object, which had repeatedly attracted my eye, but which I had as often persuaded myself was nothing more than the branchless stump of some withered tree, suddenly shifted its position, and the next moment I distinctly perceived that singular form of which the apparition had ofttimes visited my slum-

bers, but upon whose reality I now gazed for the first
time. Gliding rapidly among the trees, above the
topmost branches of many of which its graceful head
nodded like some lofty pine, all doubt was in another
moment at an end—it was the stately, the long-sought
giraffe, and putting spurs to my horse, and directing
the Hottentots to follow, I presently found myself,
half-choked with excitement, rattling at the heels of
an animal which to me had been a stranger even in
its captive state, and which thus to meet free on its
native plains has fallen to the lot of but few of the
votaries of the chase; sailing before me with incre-
dible velocity, his long swan-like neck keeping time to
the eccentric motion of his stilt-like legs, his ample
black tail curled above his back, and whisking in ludicrous concert with the rocking of his disproportioned
frame, he glided gallantly along 'like some tall ship
upon the ocean's bosom,' and seemed to leave whole
leagues behind him at each stride. The ground was
of the most treacherous description; a rotten black
soil, overgrown with long coarse grass, which concealed from view innumerable gaping fissures that
momentarily threatened to bring down my horse.
For the first five minutes I rather lost than gained
ground, and despairing over such a country of ever
diminishing the distance, or improving my acquaintance with this ogre in seven-league boots, I dis-

mounted, and the mottled carcase presenting a fair and inviting mark, I had the satisfaction of hearing two balls tell roundly upon his plank-like stern. But as well might I have fired at a wall; he neither swerved from his course nor slackened his pace, and pushed on so far ahead during the time I was reloading, that, after remounting, I had some difficulty in even keeping sight of him amongst the trees. Closing again, however, I repeated the dose on the other quarter, and spurred my horse along, ever and anon sinking to his fetlock,—the giraffe now flagging at each stride,—until, as I was coming up hand over hand, and success seemed certain, the cup was suddenly dashed from my lips, and down I came headlong —my horse having fallen into a pit, and lodged me close to an ostrich's nest, near which two of the old birds were sitting. Happily there were no bones broken, but the violence of the shock had caused the lashings of my previously broken rifle to give way, and had doubled the stock in half, the barrels only hanging to the wood by the trigger-guard. Nothing dismayed, however, by this heavy calamity, I remounted my jaded beast, and one more effort brought me ahead of my wearied victim, which stood still and allowed me to approach. In vain did I now attempt to bind my fractured rifle with a pocket-handkerchief, in order to admit of my administering the *coup de grâce.* The guard was so contracted that, as in the tantalizing phantasies of a night-mare, the hammer could not by any means be brought down upon the nipple. In vain I looked around for a stone, and sought in every pocket for my knife, with which either to strike the copper cap and bring about ignition, or hamstring the colossal but harmless animal, by whose towering side I appeared the veriest pigmy in the creation. Alas! I had lent it to the Hottentots to cut off the head of the harte-beeste, and, after a hopeless search in the remotest corners, each hand was withdrawn empty. Vainly did I then wait for the tardy and rebellious villains to come to my assistance, making the welkin ring, and my throat tingle with reiterated shouts. Not a soul appeared, and in a few minutes the giraffe, having recovered his wind, and being only slightly wounded on the hind-quarters, shuffled his long legs, twisted his bushy tail over his back, walked a few steps, then broke into a gallop, and, diving into the mazes of the forest, presently disappeared from my sight. Disappointed and annoyed at my discomfiture, I returned towards the waggons, now eight miles distant, and on my way overtook the Hottentots, who, pipe in mouth, were leisurely strolling home, with an air of total indifference as to my proceedings, having come to the conclusion that 'Sir could not fung de kameel' (catch the giraffe), for which reason they did not think it worth while to follow me, as I had directed. Two days after this catastrophe, having advanced to the Tolaan River, we again took the field, accompanied by the whole of the male inhabitants of three large kraals, in addition to those that had accompanied us from the last encampment. The country had now become undulating, extensive mimosa groves occupying all the valley, as well as the banks of the Tolaan winding among them, on its way to join the Mariqua. Before we had proceeded many hundred yards, our progress was opposed by a rhinoceros, who looked defiance, but quickly took the hints we gave him to get out of the way. Two fat elands had been pointed out at the verge of the copse the moment before, one of which Richardson disposed of with little difficulty, the other leading me through all the intricacies of the labyrinth, to a wide plain on the opposite side; on entering which, I found the fugitive was prostrate at my feet in the middle of a troop of giraffes, who stooped their long necks, astounded at the intrusion, then consulted a moment how they should best escape the impending danger, and in another were sailing away at their utmost speed. To have followed upon my then jaded horse, would have been absurd, and I was afterwards unable to recover any trace of them. . . .

"Many days elapsed before we again beheld the tall

giraffe, nor were our eyes gladdened with his sight until after we had crossed the Cashan Mountains to the country of the Baquaina, for the express purpose of seeking for him. After the many *contretemps*, how shall I describe the sensations I experienced as, on a cool November evening, after rapidly following some fresh traces in profound silence for several miles, I at length counted from the back of Breslau, my most trusty steed, no fewer than thirty-two of the various sizes, industriously stretching their peacock necks to crop the tiny leaves that fluttered above their heads in a flowering mimosa grove which beautified the scenery. My heart leapt within me, and my blood coursed like quicksilver through my veins, for, with a firm wooded plain before me, I knew they were mine; but although they stood within a hundred yards of me, having previously determined to try the *boarding* system, I reserved my fire.

"Notwithstanding that I had taken the field expressly to look for giraffes, and in consequence of several of the remarkable spoors of these animals having been seen the evening before, had taken four mounted Hottentots in my suite, all excepting Piet had, as usual, slipped off unperceived in pursuit of a troop of koodoos. Our stealthy approach was soon opposed by an ill-tempered rhinoceros, which, with her ugly old-fashioned calf, stood directly in the path, and the twinkling of her bright eyes, accompanied by a restless rolling of the body, giving earnest of her mischievous intentions, I directed Piet to salute her with a broadside, at the same time putting spurs to my horse. At the report of the gun, and sudden clattering of the hoofs, away bounded the herd in grotesque confusion, clearing the ground by a succession of frog-like leaps, and leaving me far in their rear. Twice were their towering forms concealed from view by a park of trees, which we entered almost at the same moment, and twice, on emerging from the labyrinth, did I perceive them tilting over an eminence far in advance, their sloping backs reddening in the sunshine, as with giant port they topped the ridges in right gallant style. A white turban that I wore round my hunting-cap being dragged off by a projecting bough, was instantly charged and trampled underfoot by three rhinoceroses, and long afterwards, looking over my shoulder I could perceive the ungainly brutes in the rear fagging themselves to overtake me. In the course of five minutes the fugitives arrived at a small river, the treacherous sands of which receiving their spiderlegs, their flight was greatly retarded, and by the time they had floundered to the opposite side and scrambled to the top of the bank, I could perceive that their race was run. Patting the steaming neck of my good steed, I urged him again to his utmost, and instantly found myself by the side of the herd. The lordly chief being readily distinguishable from the rest by his dark chestnut robe, and superior stature, I applied the muzzle of my rifle behind his dappled shoulder with my right hand, and drew both triggers; but he still continued to shuffle along, and being afraid of losing him should I dismount, among the extensive mimosa groves with which the landscape was now obscured, I sat in my saddle, loading and firing behind the elbow, and then placing myself across his path to obstruct his progress. Mute, dignified, and majestic stood the unfortunate victim, occasionally stooping his elastic neck towards his persecutor, the tears trickling from the lashes of his dark humid eye, as broadside after broadside was poured into his brawny front.

> " 'His drooping head sinks gradually low,
> And through his side the last drops ebbing slow
> From the red gash fall heavy one by one,
> Like the first of a thunder shower.'

Presently a convulsive shivering seized his limbs, his coat stood on end, his lofty frame began to totter, and at the seventeenth discharge from the deadly grooved bore, like a falling minaret bowing his graceful head from the skies, his proud form was prostrate in the dust. Never shall I forget the intoxicating excitement of that moment! At last, then, the summit of my ambition was actually attained, and the towering giraffe laid low! Tossing my turbanless cap into the air, alone in the wild wood, I hurra'd with bursting exultation, and unsaddling my steed, sank, exhausted with delight, beside the noble prize that I had won.

"While I leisurely contemplated the massive form before me, seeming as though it had been cast in a mould of brass, and wrapt in a hide an inch and a half in thickness, it was no longer matter of astonishment that a bullet discharged from a distance of eighty or ninety yards should have been attended with little effect upon such amazing strength.

"Two hours were passed in completing a drawing, and Piet still not making his appearance, I cut off the ample tail, which exceeded five feet in length, and was measureless the most estimable trophy I had ever gained. But on proceeding to saddle my horse, which I had left quietly grazing by the running brook, my chagrin may be conceived when I discovered that he had taken advantage of my occupation to free himself from his halter and abscond. Being ten miles from the waggons, and in a perfectly strange country, I felt convinced that the only chance of saving my pet from the clutches of the lion was to follow his trail; whilst doing which, with infinite difficulty, the ground scarcely deigning to receive a footprint, I had the satisfaction of meeting Piet and Mohanycom, who had fortunately seen and recaptured the truant. Returning to the giraffe, we all feasted merrily upon the flesh, which, though highly scented with the rank mokaala blossoms, was far from despicable, and losing our way in consequence of the twin-like resemblance of two scarped hills, we did not finally regain the waggons until after the setting sunbeams had ceased to play upon the trembling leaves of the light acacias, and the golden splendour which was sleeping upon the plain had gradually passed away."

In curious contrast to this exciting and enthusiastic but somewhat florid description, stands the very matter-of-fact account given by Mr. Roualeyn Gordon Cumming, of his first giraffe-hunt, and his *sang froid* differs amusingly from the *furore* of his fellow-Nimrod.*

Singular and striking as is the form of the giraffe, it only furnishes a proof of the wonderful manner in which an all-wise Creator has adapted means to ends. A vegetable feeder, but an inhabitant of sterile and sandy deserts, its long slender neck and sloping body enable it to reach with ease its favourite food; leaf by leaf is daintily plucked from the lofty branch by the pliant tongue, and a mouthful of tender and juicy food is speedily accumulated. The oblique and narrow apertures of the nostrils, defended even to their margins by a *chevaux-de-frise* of strong hairs, and surrounded by muscular fibres by which they can be

* A Hunter's Life in South Africa, vol. i. p. 302.

hermetically sealed, effectually prevent the entrance of the fine particles of sand which the suffocating storms of the desert raise in fiery clouds, destructive to the lord of the creation. Erect on those stilt-like legs, the giraffe surveys the wide expanse, and feeds at ease; for those mild, large eyes are so placed that it can see not only on all sides, but even behind, rendering it next to impossible for an enemy to approach undiscovered. As we reflect on these and numberless other points for admiration presented by the giraffe, we involuntarily exclaim with the Psalmist, "O Lord, how manifold are thy works; in wisdom hast thou made them all!"

> "Nature to these, without profusion kind,
> The proper organs, proper powers assigned;
> Each seeming want compensated of course,
> Here with degrees of swiftness, there of force:
> All in exact proportion to the state,
> Nothing to add, and nothing to abate."

CHAPTER VII.

RAPACIOUS BIRDS.—THE EAGLES OF ANTIQUITY.—CÆSAR'S STANDARD-BEARER.—WATERLOO.—THE DEATH OF ÆSCHYLUS.—THE MAID OF SESTOS.—THE PHŒNIX.—FLIGHT OF THE EAGLE.—GLUTTONS.—MISERABLE DEATH.—ANECDOTE.—GOLDEN EAGLE.—THE EAGLE OF WESTMINSTER.—A NOCTURNAL ALARM.—A TYRANT.—AN UNINVITED GUEST.—AN ESCAPE.—TAME EAGLE.—A USEFUL NEIGHBOUR.—A MOUNTAIN LARDER.—THE CAT-KILLER.—HUNTING IN COUPLES.—THE INVOCATION.—AMERICAN INDIANS.—EAGLES AND REINDEER.—THE GARTER.—SKUA GULL.—CREST OF THE STANLEYS.—THE EAGLE'S TEST.—THE CHILD-STEALER.—A BRAVE LAD.—A BATTLE WITH A TURBOT.—ACUTE VISION.—BALD EAGLE.—WILSON THE POET.—NIAGARA.—BENJAMIN FRANKLIN.—THE VALIANT THOMAS.—THE DEVOTED PARENTS.—AUSTRALIAN EAGLES.—THE FUTURE.

It has been happily remarked that the *Raptores* or rapacious birds, of which the eagle is the chief, and the Carnivorous animals, have a strong typical resemblance. The dispositions of both are fierce and daring; their frames, strong and sinewy, are suited alike for swift pursuit or powerful action; their sight is remarkably acute; the strong, curved, and toothed beak of the birds, like the powerful canine teeth of the *feline*, are admirably adapted for tearing; and their claws, large, curved, sharp, and retractile, are not less well fitted for holding and lacerating their struggling prey. Again, the general character of both classes is to act as a salutary check upon over-production, and to maintain that even balance in the scale of creation which is essential to the well-being of all. The birds of prey, too, like the wolves and hyænas, are of essential service in removing with rapidity dead animal matter, which, by its decomposition, would be hurtful to the living; and we find a happy adaptation to these ends in the numbers and distribution of the different species of the order, the vultures abounding in the fiery heat of the tropics, where putrefaction is most rapid; whilst the smaller falcons keep in check the myriads of lizards and other small reptiles which would otherwise be a very pest to more temperate lands.

In the Assyrian monuments, antecedent to the prophet Isaiah, the eagle is continually seen over the heads of the conquerors in battle, and was probably considered typical of victory. In the earliest of these monuments the eagle-headed human figure is one of the most prominent of the sacred types; not only is it found in colossal proportions on the walls, or guarding the portals of the chambers, but it is also constantly represented amongst the groups on the embroidered robes, and is generally seen contending with the human-headed lion or bull, of which it always appears to be the conqueror. It has been suggested that by this is intended to be denoted, the superiority of intellect over mere physical strength.

In ancient times, too, the eagle was the favourite standard of the all-conquering Romans; and in modern times, the not less adored rallying-point of the troops of Napoleon. Who can forget that scene in the history of Great Britain, upon which, perhaps, its destiny hinged; the ancient British warriors, terrible in aspect, fiercely opposing with horsemen and with their formidable chariots the landing of the invading Romans? Dismayed at the novelty of their position, and encumbered with heavy armour, the veteran troops of Cæsar shrink from the attack; a pause ensues, which is broken by a gallant warrior, —he who carried the eagle of the tenth legion,—who, first lifting his eyes to heaven, supplicates the gods to be propitious, then, with flashing eyes, and gallant mien exclaims, "Leap, leap, fellow-soldiers, unless you wish to betray your eagle to the enemy. I, for my part, will perform my duty to the commonwealth and my general!" Then, waving the eagle on high, this leader of a forlorn hope plunges into the waves and dashes towards the enemy: a tremendous shout rends the air, and one and all, burning with eagerness, the Roman soldiers leap into the sea, and struggling to the shore, join battle with the Britons; but as they can neither keep their footing nor their ranks, it would have gone hard with them had not Cæsar sent help; a fierce and bloody struggle ensues, and Britain becomes a province of Rome!

The following incident, related by Captain Siborne, in his account of the battle of Waterloo, shows the "love unto death" borne by the soldiers of Napoleon to their eagle standard. Towards the conclusion of the battle, when the Prussians had advanced in overwhelming numbers, a portion of the French had been hemmed into a churchyard. The chasseurs of the Old Guard were the last to quit the churchyard, and suffered severely as they retired. Their numbers were awfully diminished, and Pelet, collecting together about two hundred and fifty of them, found himself vigorously assailed by the Prussian cavalry from the moment he quitted the confines of Planchenoit and entered again the plain between the latter and the high-road. At one time, his ranks having opened out too much in the hurry of their retreat, some of the Prussian troops in pursuit, both cavalry and infantry, endeavoured to capture the eagle, which, covered with black crape, was carried in the midst of this devoted little band of veterans. Pelet, taking advantage of a spot of ground which afforded them some degree of cover against the fire of grape by which they were constantly assailed halted the standard-bearer and called out, "*A moi, chasseurs! Sauvons l'aigle ou mourons autour d'elle."* The chasseurs immediately pressed around him, forming what is termed the rallying square, and lowering their bayonets, succeeded in repulsing the charge of cavalry. Some guns were then brought to bear upon them, and subsequently a brisk fire of musketry, but, notwithstanding the awful sacrifice which was thus offered up in defence of their precious charge, they succeeded in reaching the main line of retreat, favoured by the universal confusion, as also by the general obscurity which now prevailed; and thus saved alike the eagle and the honour of the regiment.

There were six species of eagles known to the Romans, who entertained very fanciful notions concerning them. Of the species called *Valeria*, Pliny says, "In all the whole race of the ægles, she alone nourisheth her yong birds; for the rest, as we shall hereafter declare, doe beat them away! she only crieth not, nor keepeth a grumbling and huzzing as others doe, and evermore converseth upon the mountains." Of the species called *Boethus* he says, "Subtle she is and wittie; for when she hath seized upon tortoises, and caught them up with her tallons, she throweth them down from aloft to break their shells; and it was the fortune of the poet Æschylus to die by such a means. For when he was foretold by wizards, out of their learning, that it was his destiny to die on such a day by something falling on his head, he, thinking to prevent that, got him forth that day into a great open plain, far from house or tree, presuming upon the securitie of the clear and open skie. Howbeit an ægle let fall a tortoise, which light upon his head, dasht out his braines, and laid him asleep for ever." A sad warning this to bald-headed gentlemen! We wonder that the professors of the curling art do not make use of this tragedy to render more general the head-dress so warmly patronized by Major Pendennis. The following touching story, however, almost removes from the eagle race the stigma attaching to them for the death of Æschylus. "There hapned a marvellous example about the city Sestos, of an egle. For which in those parts there goes a great name of an egle, and highly is she honoured there. A yong maid had brought up a yong egle by hand. The egle again, to requite her kindness, would first, when she was but little, flie abroad a birding, and ever bring part of that shee had gotten unto her said nurse. In processe of time, being grown bigger and stronger, would set upon wild beasts also in the forest, and furnish her yong mistresse continually with store of venison. At length it fortuned that the damoselle died, and when her funeral pile was set a-burning, the egle flew into the midst of it, and there was consumed into ashes with the corps of the said virgin. For which cause, and in memoriall thereof, the inhabitants of Sestos and the parts there adjoining, erected in that very place a stately monument."

The perplexity of Pliny with respect to the Phœnix is highly amusing; he is evidently disposed to treat him with all respect, and to give him a place of distinction among royal birds; but in his own words, "I cannot tell what to make of him: and, first of all, whether it be a tale or no, that there is never but one of them in all the world, and the same not commonly seen. By report he is as big as an ægle; for colour as yellow and bright as gold (namely about the necke), the rest of the bodie a deep red purple; the tail azure blew, intermingled with feathers among of rose carnation colour; and the head bravely adorned with a crest and penach, finely wrought, having a tuft and plume thereupon, right faire and goodly to be seen. Manilius, the noble Roman senatour, right excellently seene in the best kind of learning and litterature, and yet never taught by any, was the first man of the long robe who wrot of this bird at large, and most exquisitely he reporteth that never man was known to see him feeding; that in Arabia he is held a sacred bird, dedicated unto the sun, that he liveth six hundred and sixty years; and when he groweth old, and begins to decay, he builds himselfe with the twigs and branches of the canel or cinnamon and frankinceuse trees, and when he hath filled it with all sorts of sweet aromaticall spices, yieldeth up his life thereupon. He saith, moreover, that of his bones and marrow there breedes at first as it were a little worme, which afterwards proveth to be a prettie bird; and the first thing that this young phœnix doth is to perform the obsequies of the former phœnix late deceased; to translate and carry away his whole nest into the citie of the Sun neere Panchea, and to bestow it full devoutly there upon the altar.... Brought he was hither to Rome in the time that Claudius Cæsar was Censor, and shewed openly to be seen in a full hall and generall assembly of the people, as appeareth upon the public records, howbeit no man ever made any doubt but he was a counterfeit phœnix and no better." It were to be wished that this paragon of a dutiful bird rested upon a more solid rock than the myths of ancient history, were it only as an example to mankind.

The eagle's flight is peculiarly expressive of strength and vigour; he wends his way with deliberate strong strokes of his powerful wing, every stroke apparently driving him on a considerable distance; and in this manner he advances through the air as rapidly as the pigeon or any other bird which may appear to fly much more quickly: velocity of flight, it may be remarked, generally depends upon the rapidity with which the strokes of the wings succeed each other; a simple downward stroke would only tend to raise the bird in the air. To carry it forwards the wings require to be moved in an oblique plane, so as to strike backwards as well as downwards: the turning in flight is principally effected by an inequality in the vibration of the wings. The rapidity with which a strong bird of prey flies in pursuit of his quarry is inconceivably great; the flight of a hawk is calculated at one hundred and fifty miles an hour, and the anecdote of the falcon belonging to Henry IV. of France, which flew in one day from Fontainebleau to Malta, a distance of thirteen hundred and fifty miles, is well authenticated.

Notwithstanding the facility with which he flies, when once fairly launched, a very slight wound disables the eagle from rising into the air when on level ground. Even after having gorged himself to excess (and there is no greater glutton than this king of the air), the eagle is unable to rise, and falls a victim occasionally to his want of moderation in feeding. In Sutherlandshire, Mr. St. John twice fell in with instances of eagles being knocked down when unable to fly from over-eating. A dumb, eccentric character killed one with a stick; and in the other instance a shepherd-boy found an eagle gorging itself on some drowned sheep in a watercourse, and being, like all herd-boys, as skilful as David in the use of sling and stone, he broke the eagle's pinion with a pebble, and actually stoned the poor bird to death. In this case the eagle was taken at peculiar disadvantage, being surprised in a deep rocky burn, out of which he would have had difficulty in rising quickly even if he had not dined so abundantly.

An eagle had been caught in a vermin-trap, and by his struggles had drawn the peg by which the trap was fastened to the ground, and had flown away with it. Nothing was seen for some weeks of eagle or trap, till one day a sportsman seeing some strange object hanging from a branch of a tree, went to examine what it was, and found the bird hanging by his leg, which was firmly held by the trap. The chain and peg had got fixed among the branches, and the poor bird had died miserably from starvation suspended by the foot.

In Scotland, advantage is taken of the difficulty attending the rise of the eagle from level ground to catch him in the following manner. Some desert place frequented by eagles is selected, and four walls are built like those of a hut, an opening being left at the foot, large enough to allow of the bird's walking in and out. To the outside of this opening a strong cord with a running noose is fixed; all being so arranged, a dead sheep or other carrion is thrown into the enclosure. This is eagerly attacked by the eagle, who gorges himself to excess, and becomes half stupefied: he does not attempt to rise into the air, but walks out of the opening; the running noose soon catches him round the neck, as a hare is caught in a springe, and his own struggles do the rest.

Some years ago, in Nottinghamshire, a groom was exercising a horse in the early morning, when a terrier which was with him put up from a bush a magnificent eagle, which flew slowly over the hedge into the adjoining field, pursued by the dog, who came up with and attacked it before it could fairly rise; a fierce struggle ensued, but the dog, though severely bitten, maintained his hold, and the bird, which measured eight feet across the wings, was eventually secured. He, too, was captured from having over-indulged in the luxury of carrion.

The late amiable Bishop of Norwich, whose enthusiasm as a naturalist is well known, gives from observation the following account of a golden eagle as seen in his native wilds. Whilst climbing some precipices, near a great waterfall, in the volcanic district of Auvergne, there arose above the roar of the waters a short shrill cry, coming, as it were, from the clouds; on looking in the direction whence it came, a small dark speck was seen moving steadily onwards: it was a golden eagle evidently coming from the plain countries below. As he came nearer, he seemed to float or sail in mid-air, only occasionally gentle flapping his wings as if to steady him. Though when first seen he was at the distance of a full mile, in less than a minute he was within gun-shot, and the observer having concealed himself, the bird looked round once or twice, darted down his legs, and alighted on a rock within a few yards of him. For a moment the eagle gazed a out with his sharp bright eyes, as if to assure himself that all was safe, then for a few moments more, nestled his head beneath one of his expanded wings, and appeared to plume himself. Having done this, he stretched out his neck and looked keenly and wistfully towards the quarter of the heavens whence he came, and uttered a few rapid screams; then stamping with his feet, he protruded his long hooked talons, at the same time snapping his beak with a sharp noise like the cracking of a whip. There he remained for about ten minutes, manifesting great restlessness, when suddenly he seemed to hear or see something, and immediately rising from the rock, floated away to meet his mate, now seen approaching. After soaring in a circle, they went away, and were no more seen.

The Chartists and special constables of Westminster, who were preparing on the 9th of April, 1848, for the grand "Demonstration" of the following day, beheld with varied feelings an omen which they interpreted according to their views. A magnificent eagle suddenly appeared sailing over the towers of Westminster Abbey, and after performing numerous gyrations, was seen to perch upon one of the pinnacles of the abbey. He formed a most striking object, and a crowd speedily collected to behold this unusual spectacle. After gazing about him for a time, he rose, and began ascending by successive circles to an immense height, and then floated off to the north of London, occasionally giving a gentle flap with his wings, but otherwise appearing to sail away to the clouds, among which he was ultimately lost.

Whence came this royal bird, and whither did he wend his way?

His history was as follows. Early in 1848 a white-tailed sea-eagle was brought to London in a Scotch steamer, cooped up in a crib used for wine-bottles, and presenting a most melancholy and forlorn appearance. A kind-hearted gentleman seeing him in this woful plight, took pity on him, purchased him, and took him to Oxford, he being duly labelled at the Great Western Station, "Passenger's Luggage." By the care of his new master, Mr. Francis Buckland, the bird soon regained his natural noble aspect, delighting especially to dip and wash in a pan of water, then sitting on his perch with his magnificent wings expanded to their full extent, basking in the sun, his head always turned towards that luminary, whose glare

R

he did not mind. A few nights after his arrival at his new abode, the whole house was aroused by cries as of a child in mortal agony. The night was intensely dark, but at length the boldest of the family ventured out to see what was the matter. In the middle of the grass-plot was the eagle, who had evidently a victim over which he was cowering with outspread wings, croaking a hoarse defiance to the intruder upon his nocturnal banquet. On lights being brought, he hopped off with his prey in one claw to a dark corner, where he was left to enjoy it in peace, as it was evidently not an infant rustic from the neighbouring village, as at first feared. The mystery was not, however, cleared up for three days, when a large lump of hedgehog's bristles and bones rejected by the bird at once explained the nature of his meal. He had doubtless caught the unlucky hedge-pig—as it is called in Oxfordshire—when on his rounds in search of food, and in spite of his formidable armour of bristles, had managed to uncoil him with his sharp bill, and to devour him. How the prickles found their way down his throat, is best known to himself; but it must have been rather a stimulating feast.

This eagle was with good reason the terror of all the other pets of the house. On one occasion he pursued a little black and tan terrier, hopping with fearful jumps, assisted by his wings, which, happily for the affrighted dog, had been recently clipped. To this the little favourite owed his life, as he crept through a hedge which his assailant could not fly over; but it was a very near thing, as, if the dog's tail had not been between his legs, it would certainly have been seized by the claw which was thrust after him just as he bolted through the briars. Less fortunate was a beautiful little kitten, the pet of the nursery,—a few tufts of fur alone marked the depository of her remains. Several guinea-pigs and sundry hungry cats, too, paid the debt of nature through his means; but a sad loss was that of a jackdaw of remarkable colloquial powers and unbounded assurance, who, rashly paying a visit of a friendly nature to the eagle, was instantly devoured. Master Jacko, the monkey, on one occasion only saved his dear life by swiftness of foot, getting on the branch of a tree just as the eagle came rushing to its foot with outspread wings and open beak. The legend is, that Jacko became rather suddenly grey after this; but the matter is open to doubt.

One fine summer's morning the window of the breakfast-room was thrown open previous to the appearance of the family. On the table was placed a ham of remarkable flavour and general popularity, fully meriting the high encomiums which had been passed upon it the previous day. The rustling of female garments was heard, the breakfast-room door opened, and—oh, gracious! what a sight! There was the eagle perched upon the ham, tearing away at it with unbounded appetite, his talons firmly fixed in the rich deep fat. Finding himself disturbed, he endeavoured to fly off with the prize, and made a sad clatter with it among the cups and saucers; finding, however, that it was too heavy for him, he suddenly dropped it on the rich carpet, snatched up a cold partridge, and made a hasty exit through the window, well satisfied with his foraging expedition. The ham, however, was left in too deplorable a state to bear description. The eagle was afterwards taken to London and placed in a courtyard near Westminster Abbey, where he resided in solitary majesty. It was from thence he made his escape on the 9th of April. He first managed to flutter up to the top of the wall; thence he took flight unsteadily and with difficulty until he had cleared the houses, but as he ascended into mid-air his strength returned, and he soared majestically up, as has been narrated. After his disappearance, his worthy master said with a disconsolate air: "Well, I've seen the last of my eagle!" but thinking that he might possibly find his way back to his old haunt, a chicken was tied to a stick in the court-yard, and just before dark, master eagle came back, his huge wings rustling in the air. The chicken cowered down to the ground, but in vain—the eagle saw him, and pounced down in a moment in his old abode. Whilst he was busily engaged in devouring the chicken, a plaid was thrown over his head, and he was easily secured. After this escapade he was sent to the Zoological Gardens, Regent's Park, where he may be recognized by his having lost the outside claw of the left foot.*

Fierce and wild as the golden eagle generally is, instances have occurred in which it has been thoroughly tamed. Captain Green, of Buckden, in Huntingdonshire, had in his possession a splendid bird of this description, which he had himself trained to take hares and rabbits. Another instance is known of an eagle captured in Ireland after it had attained maturity, which speedily became domesticated and firmly attached to the place where it was fed, to which it always returned though perfectly at liberty. Its wings had, indeed, been cut when first brought thither, but they were allowed to grow again; and this magnificent bird, on recovering the use of them, would repeatedly soar away and absent himself for a fortnight or three weeks. It became very much attached to those who were in the habit of feeding or caressing it. On its first arrival it had been placed

* Further interesting particulars of this noble bird are to be found in that excellent and original work by Mr. Buckland, 'Curiosities of Natural History,' second series, p. 101.

in a garden situated in a slope overhanging a lake; a shed had been built for its accommodation, but it generally preferred a perch of its own selection— the branch of a large apple-tree which grew out nearly in a horizontal position from the stem. Its food was chiefly crows, which were shot for it; sometimes it attempted to procure them for itself, but never successfully, as their agility, in turning short and rapidly, enabled them to elude its superior strength of wing; latterly, therefore, it contented itself with eyeing them wistfully as they flew or perched securely over its head. It was never suspected of committing any havoc among the sheep or lambs in the adjoining fields, but now and then, when from some accident it had not been regularly supplied with its accustomed food, it would seize upon and kill young pigs. Children, who constantly met it as it walked about the garden, were never molested; but on one occasion it attacked its master with some violence, in consequence, it was supposed, of his having neglected to bring some bread or other food it was accustomed to receive from his hand. At length, after having lived nearly twelve years in this way, this interesting bird was killed by a ferocious mastiff; no one saw the battle, but it must have been long and bravely contested, for the dog, though victorious, was so severely wounded that he died almost immediately afterwards.

Until the young eagles are fully able to fly and maintain themselves, the old birds keep them supplied with provisions most abundantly. Smith, in his 'History of Kerry,' relates that a poor man in that county got a comfortable subsistence for his family, during a famine, by robbing an eagle's nest of the food brought for the eaglets, whose period of helplessness he protracted by clipping their wings; but the most curious account of one of these eagle-nest larders is related by a gentleman who was visiting at a friend's house in Scotland, near which was a nest, which for several summers two eagles had occupied. It was on a rock, and within a few yards of it was a stone about six feet long, and nearly as broad, and upon this stone almost constantly, but always when they had young, there were to be found, grouse, partridges, hares, rabbits, ducks, snipes, rats, mice, and sometimes kids, fawns, and lambs. When the eaglets were able to hop the length of this stone, the eagles often brought hares and rabbits alive, and placing them before their young, taught them to kill and tear them to pieces, just as a cat teaches its kittens to kill mice. Sometimes, it seems, the hares got off from the young ones whilst they were amusing themselves with them, and one day a rabbit escaped into a hole where the old eagle could not find it. Another day, a young fox-cub was brought, which, after it had fought well, and desperately bitten the young ones, attempted to make its escape up the hill, and would probably have succeeded, had not a shepherd, who was watching the motions of the eagles with a view to shoot them, prevented it. As the eagles kept what might be called such an excellent storehouse, the gentleman said that, whenever visitors came unexpectedly, he was in the habit of sending his servants to see what his neighbours the eagles had to spare, and that they scarcely ever returned without some dainty dishes, all the better for being rather high.

When the hen-eagle was hatching, the table was kept well furnished for her use, and her attentive mate generally tore a wing from a bird, or a leg from a hare, with which he supplied her. These birds were very faithful to one another, and would never permit even their young to build anywhere near them.

The marten and wild cat are favourite morsels with eagles. A tame one which Mr. St. John kept for some time, killed all the cats about the place. Sitting motionless on his perch, he waited quietly, and seemingly unheedingly, till the unfortunate animal came within reach of his chain; then down he flew, and enveloping the cat with his wings, seized her in his powerful talons, with one foot planted firmly on her loins, and the other on her throat, and nothing more was seen of poor Grimalkin, except her skin, which the eagle left empty and turned inside out, like a rabbit-skin hung up by the cook; the whole of the carcase, bones and all, being stowed away in the bird's capacious maw.

Mr. Thompson, an eminent naturalist of Ireland, was once out hunting among the Belfast mountains, when suddenly an eagle appeared above the hounds as they came to fault on the ascent to Devis; presently they came on the scent again, and were in full cry, the eagle hovering above them, when suddenly he dashed forward, and carried off the hare from under the very noses of the dogs. Mr. St. John has seen an eagle pounce on a pack of grouse, and with outspread wings so puzzle and confuse them, that he seized and made off with two or three, before the others, or, indeed, the sportsmen, recovered from their astonishment. The golden eagle has been seen in Sicily to hunt in couples; one of the birds would make a loud rustling by a violent beating of its wings against bushes and shrubs, whilst the other remained in ambush at a short distance, watching for anything that might appear; if a rabbit or hare was driven out, it was immediately pounced upon, and the prey thus obtained was shared between the depredators.

Eagles are said to be very long-lived. One that died in Vienna was stated to have lived in confinement one hundred and four years. From the great value attached by the North American Indians to an eagle's plume, which is considered equivalent in value to a fine horse, their hunters are continually on the lookout to catch or to kill these birds. Sometimes a hole is dug, and slightly covered, and, there buried as it were, an Indian will remain for days together, with a bird on his hand as a lure for the eagle; at other times, the carcase of a deer is displayed, and the indefatigable hunter will watch, rifle in hand, with equal patience in some neighbouring place of concealment, until his perseverance is rewarded with success.

A story is current on the plains of Saskatchewan, of a half-bred Indian, who was vaunting his prowess before a band of his countrymen, and wished to impress them with a belief of his supernatural and necromantic powers. In the midst of his florid harangue an eagle was observed suspended in the air directly over his head, upon which, pointing aloft with his dagger, which glistened brightly in the sun, he called upon the royal bird to come down. To his utter amazement, and to the consternation of the surrounding Indians, the eagle seemed to obey the charm, for instantly shooting down with the velocity of an arrow, he impaled himself on the point of the glittering weapon, which had, of course, been the object of attraction.

The distinguishing mark of a chieftain in the Highland clans was an eagle's feather in the bonnet; and among the North American Indians, the same ornament is esteemed in the highest degree. The young Indian "brave" glories in his eagle's plume, as the emblem of might and courage, and regards it as the most honourable decoration with which he can adorn himself. In 1734, Tomochichi, King of the Yammacrows, and several other Indian chiefs, arrived in England, and were introduced to George II., at Kensington; on that occasion, Tomochichi presented to his Majesty a gift of eagles' plumes, being the most respectful gift he could offer, and concluded an eloquent speech in these words,—"These are the feathers of the eagle, which is the swiftest of birds, and who flieth all round our nations. These feathers are a sign of peace in our land, and we have brought them over to leave them with you, O great king, as a sign of everlasting peace."

The eagle-feathers are also attached to the calumets, or smoking-pipes, used in the celebration of their most solemn festivals, hence the bird has obtained the name of the "Calumet Eagle."

In some parts eagles play sad havoc with the young lambs, and occasionally with the herds also. That there is foundation for the following statement, made by Mr. Regnard, there can be no doubt; but, like the old tale of the gigantic Patagonians, it has not lost by repetition. The worthy traveller says, "There are also some birds which carry on a destructive warfare with the reindeer, and among the rest, the eagle is extremely fond of the flesh of this animal. In this country, great numbers of eagles are to be found, of such an astonishing size, that they often seize upon with their claws the young reindeer of three or four months old, and lift them up in this manner to their nests, at the tops of the highest trees. This particular immediately appeared to me very doubtful, but so true is it that the guard employed to watch the young reindeer, is only used for this very purpose. All the Laplanders have given me the same information; and the Frenchman, who was our interpreter, assured me that he had seen many examples of it, and that having one day followed an eagle which carried a young reindeer from its mother's side to its own nest, he cut the tree at the foot, and that the half of the animal had already been eaten by the young ones. He seized the young eagles, and made the same use of them which they had made of his young deer, namely, he ate them: their flesh was pretty good, but black, and somewhat insipid."

It would seem that in the Orkney Islands there were persons who professed to have the power, by means of a sort of incantation, of causing the plunderers to abandon their spoils; of which the following amusing account is given by Brand, who visited the Orkneys towards the end of the seventeenth century.

"There are," says that writer, "many eagles which destroy their lambs, fowls, etc., for the preventing of which, some, when they see the eagles catching, or fleeing away with their prey, use a charm by taking a string whereon they cast some knots, and repeat a form of words, which being done, the eagle lets her prey fall, though at a great distance from the charmer, an instance of which I had from a minister, who told me that, about a month before we came to Zetland, there was an eagle that flew up with a cock at Scalloway, which one of these charmers seeing, presently took a string (her garter, it is supposed), and casting some knots thereupon, with using the ordinary words the eagle did let the cock fall into the sea, which was recovered by a boat that went out for that end."

In the Shetlands the skua gull is held in particular regard by the natives, as from the inveterate hostility borne by them to the eagle and raven, the great enemies of the lambs, they serve as valuable protectors to these defenceless animals. No sooner does the eagle emerge from his eyrie amid the cliffs than the skua descend upon him in bodies of three and four, and soon cause him to beat a precipitate retreat. An eye-witness describes such a scene: an eagle was returning to his eyrie in the western crags of Foula, and, contrary to his usual wary custom, was making a short cut by crossing an angle of land; not a bird was discernible, but suddenly the majestic flight of the eagle ceased, and he descended hurriedly, as if in the act, of pouncing; in a moment five or six of the skua cleft the air with astonishing velocity; their wings were partly closed and perfectly steady, and as they thus shot through the air they soon came up with the eagle, and a desperate engagement ensued. The skua never ventured to attack the enemy in front, but taking a short circle around him, until his head and tail were in a direct line, the gull made a desperate stoop, and striking the eagle on the back, darted up again almost perpendicularly, and fell to the rear. Three or four of these birds passing in quick succession, harassed the eagle most unmercifully; the engagement continued to the decided disadvantage of the eagle, till the whole were lost in the rocks.

There are many instances on record of infants being carried away by the larger birds of prey, and, in fact, there is scarcely a district infested by them which has not some tale of the sort. The following is contained in the first volume of the 'Monasticon Anglicanum,' and may possibly have been founded on fact, though probably embellished by the ancient chronicler:—

"Alfred, King of the West Saxons, went out one

day a-hunting, and passing by a certain wood, heard,
as he supposed, the cry of an infant from the top of a
tree, and forthwith diligently inquired of the huntsmen
what that doleful sound could be, and commanded one
of them to climb the tree; when in the top of it was
found an eagle's nest, and, lo! therein a pretty sweet-
faced infant, wrapped up in a purple mantle, and upon
each arm a bracelet of gold, a clear sign that he was
born of noble parents. Whereupon the king took
charge of him, and caused him to be baptized, and
because he was found in a nest, he gave him the
name of Nestingum, and in after-time, having nobly
educated him, he advanced him to the dignity of an
earl."

The crest of the Stanley family is an eagle preying
upon a child, the origin of which is said by Dugdale,
in his 'Baronage of England,' to be as follows:—

"A certain Thomas de Lathom had an illegitimate
son, called Oskytel, and having no child by his own
lady, he designed to adopt this Oskytel for his heir,
but so that he himself might not be suspected for his
father. Observing, therefore, that an eagle had built
her nest in a large spread oak within his park at
Lathom, he caused the child, in swaddling clothes, to
be privily conveyed thither; and (as a wonder) presently called forth his wife to see it, representing to
her that having no male issue, God Almighty had

thus sent him a male child, and so preserved that he looked upon it as a miracle, disguising the truth so artificially from her that she forthwith took him with great fondness into the house, educating him with no less affection than if she had been his natural mother, whereupon he became heir to that fair inheritance. And that in token thereof not only his descendants, while the male line endured, but the Stanleys proceeding from the said Isabel, have ever since borne the *child in the eagle's nest with the eagle thereon* for their crest."

This, by the way, recalls to our mind a curious passage in that venerable book, Guillim's 'Display of Heraldry,' which alludes to a circumstance probably known to few of our readers:—

"It is related that the old eagles make a proof of their young by exposing them against the sunbeams, and such as cannot steadily behold the brightness are cast forth as unworthy to be acknowledged their offsprings. In which respect William Rufus, king of this land, gave for his device an eagle looking against the sun, with this word, *perfero* (I endure it), to signify that he was not in the least degenerated from his puissant father the Conqueror."

A deplorable circumstance occurred in Sweden, which has become matter of tradition from its melancholy interest. A young and blooming mother, whilst occupied in the fields, had laid her first-born, the pride of her heart, on the ground at a short distance from her; the babe was tranquilly sleeping, when suddenly a huge eagle swooped down and carried him off in his talons. In vain the mother pursued with frantic cries; in vain she implored aid from others; for a considerable time the screams of the poor infant were heard, but they gradually became fainter and fainter in the distance, and the wretched mother saw her child no more. The shock was too much—her reason left its seat, and she, the pride and ornament of the village, became the inmate of a lunatic asylum! On a high-pointed pinnacle of inaccessible rock near the summit of the Jungfrau, one of the loftiest Alps, there were long to be seen fluttering in the breeze the tattered remains of the clothing of an infant which had been carried thither and leisurely devoured by a læmmergeyer.

It is satisfactory, however, to find that in some instances these fierce marauders are punished for their temerity, of which a striking example occurred in the parish of St. Ambrose, near New York. Two boys, aged respectively seven and five, were amusing themselves by trying to reap while their parents were at dinner. A large eagle soon came sailing over them, and with a sudden swoop, attempted to seize the eldest, but missed his aim; the bird, not at all dismayed, alighted at a short distance, and in a few moments repeated the attack; the bold little fellow, however, gallantly defended himself with the sickle, and when the bird rushed at him, resolutely struck at it; the sickle entered under the wing, went through the ribs, and laid the bird dead. On opening its stomach it was found entirely empty, which may explain such an unusually bold attack.

A gamekeeper was on the moors in Scotland, when he observed an eagle rise from the ground with something he had seized as his prey; for a time he flew steadily, but suddenly became agitated, fluttered, spired upwards in a straight line to a vast height, then ceasing to flap his wings, he fell headlong to the ground. Struck with so unaccountable an occurrence, the man hastened to the spot, and found the eagle quite dead, with a wounded stoat struggling by his side; the stoat, when in the air, had fixed himself on his assailant's throat, and completely turned the tables on him.

Eagles, if they can take a fine fish at a disadvantage, will not hesitate to vary their diet; but unexpected difficulties sometimes arise and prevent their enjoyment of the little treat, of which a pleasant story is told by Brand, as having happened off the Orkney Islands.

"About six years since an eagle fell down on a

turbot sleeping on the surface of the water, on the east side of Brassa; and having fastened his claws in her, he attempted to fly up; but the turbot awakening, and being too heavy for him to fly up with, endeavoured to draw him down beneath the water. Thus they struggled for some time, the eagle labouring to go up, and the turbot to go down, till a boat that was near to them and beheld the sport, took them both, selling the eagle to the Hollanders then in the country."

An instance of the boldness of eagles is mentioned by Mr. Lear, in his very interesting 'Journal of a Landscape Painter.' When sketching the formidable fortress of Khimára, in Albania, there came two old women with the hope of selling some fowls, which they incautiously left on a ledge of rock just above their heads, whilst they discussed the terms of the purchase with Anastásio, Mr. Lear's dragoman. When behold! two superb eagles suddenly floated over the abyss, and—pounce—carried off each his hen; the unlucky *gallinaceæ* screaming vainly as they were transported by unwelcome wings to the inaccessible crags on the far side of the ravine, where young eagles and destiny awaited them.

Near Joánnina, Mr. Lear saw jays and storks and vultures in vast numbers. Owing to a disease among the lambs, the birds of prey had gathered together, and a constant stream of these harpies passed from the low grounds to the rocks above. One hundred and sixty were counted on one spot; and as with outstretched necks and wings they soared and wheeled, their appearance was very grand.

One of the most surprising facts connected with birds of prey is that wonderful acuteness of vision which enables the eagle, for example, when soaring in the clouds, to discern, and to pounce with unerring precision on so small an object as a grouse upon the ground. When looking for its prey, the eagle sails in large circles, with tail outspread, and wings scarcely moving. Thus it soars aloft in a spiral course, its gyrations becoming less and less perceptible until it dwindles to a mere speck, and is at length lost to view; when suddenly it reappears, rushing down like lightning, and carries off in its talons some unhappy prey.

The raptorial birds are endowed with a very beautiful modification of the eye in relation to this power of vision; the globe is surrounded with a circle of bony plates, slightly moving on each other, whereby its form is maintained, and the muscles at the back of the eye are so arranged that by their pressure, the front of that organ can be rendered more prominent than is ever seen in Mammalia, or they can be quite relaxed, and the front of the eye rendered nearly flat. The first condition fits it for discerning near objects, the second endows it with telescopic sight, by the peculiar adaptation of the refractive media, and is that which exists when the bird is hovering on high.

The Bald eagle, the emblem of America, is remarkable for his great partiality to fish, and his superior strength enables him to turn the industry of the osprey to his own account, by robbing it of its prey. The following spirited description of such a scene is from the pen of the poet-naturalist, Wilson:—"Elevated on the high dead limb of some gigantic tree, that commands a wide view of the neighbouring shore and ocean, he seems calmly to contemplate the motions of the various feathered tribes that pursue their busy avocations below: the snow-white gulls slowly winnowing the air; the busy tringæ coursing along the sands; trains of ducks streaming over the surface; silent and watchful cranes, intent and wading; clamorous crows and all the winged multitudes that subsist by the bounty of this vast liquid magazine of Nature. High over all these hovers one whose action instantly arrests all his attention. By his wide curvature of wing, and sudden suspension in the air, he knows him to be the fish-hawk, settling over some devoted victim of the deep. His eye kindles at the sight, and balancing himself with half-opened wings on the branch, he watches the result. Down, rapid as an arrow from Heaven, descends the object of his attention; the roar of its wings reaching the ear as it disappears in the deep, making the surges foam around: at this moment the eager look of the eagle is all ardour, and, levelling his neck for flight, he sees the fish-hawk once more emerge, struggling with his prey, and mounting in the air with screams of exultation. These are the signal for our hero, who, launching into the air, instantly gives chase, and soon gains on the fish-hawk: each exerts his utmost to mount above the other, displaying in these rencontres the most sublime aerial evolutions. The unencumbered eagle rapidly advances, and is just on the point of reaching his opponent, when, with a sudden scream, probably of despair and honest execration, the latter drops his fish: the eagle, poising himself for a moment, as if to take a more certain aim, descends like a whirlwind, snatches it in its grasp ere it reaches the water, and bears his ill-gotten booty silently away to the woods."

The awful gulf into which the waters tumble at the Horse-shoe Fall at Niagara, is a favourite resort of the eagles. They may be seen sailing about in the mist which rises from the turbulent waters, with an ease and elegance of motion almost sublime.

"High o'er the watery uproar silent seen,
Sailing sedate in majesty serene,
Now 'midst the pillared spray sublimely lost,
And now emerging, down the rapids tost,

> Glides the bald eagle, gazing calm and slow
> O'er all the horrors of the scene below,
> Intent alone to sate himself with blood
> From the torn victims of the raging flood."

The attraction that leads these birds to the Falls, is the swollen carcases swept down the river, and precipitated over the cataract. Wilson saw an eagle seated on a dead horse, keeping a whole flock of vultures at a distance till he had satisfied himself; and on another occasion, when thousands of tree-squirrels had been drowned in their migration across the Ohio, and hosts of vultures had collected, the sudden appearance of a bald eagle sent them all off, and the eagle kept sole possession for many days.

Notwithstanding the poetical description we have quoted from the pen of Wilson, it is to be feared that the bald eagle is but a reprobate, and too well deserves the following character, given to it by the celebrated Benjamin Franklin:—

"For my part (says he) I wish the bald eagle had not been chosen as the representative of our country. He is a bird of bad moral character; he does not get his living honestly. You may have seen him perched on some dead tree, where, too lazy to fish for himself, he watches the labours of the fishing-hawk; and when that diligent bird has at length taken a fish, and is bearing it to his nest for the support of its mate and young ones, the bald eagle pursues him and takes it from him. With all this injustice he is never in good case, but, like those among men who live by sharping and robbing, he is generally poor, and often very lousy. Besides, he is a rank coward; and the little king bird, not bigger than a sparrow, attacks him boldly and drives him out of the district. He is, therefore, by no means a proper emblem for the brave and honest Cincinnati of America, who have driven all the king birds from our country; though exactly fit for that order of knights which the French call *chevaliers d'industrie.*"

These bald eagles are, indeed, sad thieves, not confining themselves to fish or such small deer. Mr. Gardiner, of Long Island, saw one flying with a lamb ten days old, and by hallooing and gesticulating caused the bird to drop it, but the back was broken. The same gentleman shot one seven feet from tip to tip of the wings, which was so fierce, that when attacked by a dog it fastened its claws into his head, and was with difficulty disengaged. Another case is on record, when one of these eagles pounced upon a strong tomcat and flew away, but puss offered such a vigorous resistance with his teeth and claws, that a regular battle took place in the air; at length, tired of struggling, and extremely incommoded by the claws of the cat, the eagle descended to the earth, where the battle continued, but was terminated by some men who captured both combatants, much the worse for wear.

In New Jersey, a woman weeding in her garden had set her child down near, when a sudden rush and a scream from the infant alarmed her, and starting up, she beheld the child being dragged along the ground by a huge eagle. Happily the frock, in which the bird's talons were fixed, gave way, and alarmed by the outcry of the mother, he did not offer to renew the attack, but flew away.

The chief redeeming feature in the character of the bald eagle is its love for its young. During the process of clearing a piece of land, fire was set to a large dead pine-tree, in which was an eagle's nest and young; the tree being on fire more than halfway up, and the flames rapidly ascending, the parent eagle darted around and among them until her plumage was so much scorched that it was with difficulty she could escape, and even then attempted several times to return to her offspring's assistance.

Dr. Richardson relates an adventure which befell him, showing the determination with which the gyrfalcon will also defend its offspring. "In the middle of June, 1821," says he, "a pair of these birds attacked me, as I was climbing in the vicinity of their nest, which was built on a lofty precipice on the borders of Point Lake. They flew in circles, uttering loud and harsh screams, and alternately stooping with such velocity that their motion through the air produced a loud rushing noise. They struck their claws within an inch or two of my head, and I endeavoured, by keeping the barrel of my gun close to my cheek, and suddenly elevating the muzzle when they were in the act of striking, to ascertain whether they had the power of instantaneously changing the direction of their rapid course, and found that they invariably rose above the obstacle with the quickness of thought, showing equal acuteness of vision and power of motion."

To those who know how low in the scale of intelligence the Marsupial animals rank, it is not very flattering to the dignity of the lords of the creation to find that we have been at times confounded with them, even by so keen-sighted and quick-witted a bird as an eagle. An anecdote related by Captain Flinders is an amusing illustration of such a blunder, which must, by the way, have sorely perplexed the birds; the scene is laid on "Thistle's Island."

"In our way up the hills to take a commanding station for the survey, a speckled yellow snake lay asleep before us. By pressing the butt-end of a musket on his neck I kept him down, whilst Mr. Thistle, with a sail-needle and twine, sewed up his mouth, and he was taken on board alive for the naturalist to examine. We were proceeding onward with our prize when a white eagle, with fierce aspect and outspread wing, was seen bounding towards us, but stopping short at twenty yards off, he flew up into a tree. Another bird of the same kind discovered himself by making a motion to pounce down upon us as we passed underneath; and it seemed evident they took us for kangaroos, having probably never before seen an upright animal in the island of any other species. These birds sit watching in the trees, and should a kangaroo come out to feed in the day-time, it is seized and torn to pieces by these voracious creatures."*

The following lines, by Southey, elegantly advert to a myth of the ancients, which obscurely shadows forth that transition which human nature is destined to undergo in our progress from one condition of existence to another. Like the bird, we shall leave behind us in this world all that is gross, impure, and perishable; and as she is fabled to rise from the waters, so we hope to rise from the earth, purified, glorified, and immortal.

"Even as the eagle (ancient storyers say),
When faint with years she feels her flagging wing,
Soars up towards the mid-sun's piercing ray.
Then, filled with fire, into some living spring
Plunges, and casting there her ancient plumes,
The vigorous strength of primal youth resumes."

* "A Voyage to Terra Australis, vol. i. p. 188."

CHAPTER VIII.

SPANIARDS IN MEXICO.—ANCIENT MEXICANS.—FEATHER EMBROIDERY.—MONTEZUMA'S AVIARY.—GORGEOUS ARRAY.—WORKS OF ART.—DISTRIBUTION OF HUMMING-BIRDS.—RAPIDITY OF FLIGHT.—NESTS.—COURTING.—SINGULAR BOWER.—ACTIONS OF POLYTMUS.—NEST-MAKING.—ANECDOTE.—MODE OF CAPTURE.—WITTY EPITAPH.—THE PHILOSOPHER AND THE MIDDIES.—PETS.—TROCHILUS IN CAPTIVITY.—SONG.—BLUEFIELD'S RIDGE.—GORGEOUS SCENE.—PUGNACITY OF HUMMING BIRDS.—A COMBAT.—THE "DOCTOR BIRD."—FAVOURITE RESORT.—TAMENESS.—A BOLD BIRD.—INTERESTING BIRDS.—NOTES FOR ORNITHOLOGISTS.—TONGUE OF HUMMING-BIRD.—BIRDS AND SPIDERS.—MODE OF TAKING FOOD.—MR. GOULD'S COLLECTION.—ELEGANT ARRANGEMENT.—"THE TROCHILIDÆ."—DOMESTICATION.

" When the morning dawns, and the blest sun again
  Lifts his red glories from the eastern main ;
  Then, through our woodbines, wet with glittering dews,
  The flower-fed Humming-bird his round pursues ;
  Sips, with inserted tube, the honeyed blooms,
  And chirps his gratitude as round he roams ;
  While richest roses, though in crimson drest,
  Shrink from the splendour of his gorgeous breast.
  What heavenly tints in mingling radiance fly!
  Each rapid movement gives a different dye.
  Like scales of burnished gold they dazzling show—
  Now sink to shade—now like a furnace glow."—WILSON.

STERN, bigoted, and cruel, were those fierce rapacious men, the Spanish conquerors of Mexico : men cast in an iron mould, which rendered them insensible to all ordinary emotions. It is, however, recorded of Cortez and his companions, that as, on their route to Cempoalla, they marched through a wilderness of noble trees, from whose branches the most beautiful blossoms were suspended, and trod underfoot wild roses, honeysuckles, and sweet-smelling herbs, expressions of admiration escaped them ; and when, in addition to these charms of vegetation, clouds of gorgeous butterflies arose, and birds of glorious plumage filled the air with delicious melody, the apathy of these warriors was completely overcome, and they involuntarily burst forth in exclamations of delight, terming the country a terrestrial paradise, and fondly comparing it to the fairest regions of their own sunny land.

First in beauty among those birds which struck them with admiration were the *Tomineios*, or Humming-birds, which, as old Herrara says, they doubted whether they were bees or butterflies ; and civilized man has since vied with the Indian in inventing expressions of admiration of these fair objects. But here—as on other occasions—the child of nature has proved the better poet, and no term has been invented more expressive than their Indian name *Guarocigaba*, which signifies the *Beams or Locks of the Sun*. Before this, the *cheveux de l'astre de jour* of Buffon is a tame comparison.

It is an interesting fact, that, as a general rule birds of the most brilliant plumage are found in those parts of the world where the sun shines brightest, the flowers are the loveliest, and where gems and precious metals abound; as if Nature had bountifully brought together the objects most attractive to man. The rubies and the emeralds of the earth are, however, cast into the shade by the living gems which float in the air above them.

Holding a sort of analogy to the mosaic-work of the Italians, and, like it, standing unrivalled, was the the wonderful featherwork of the ancient Mexicans. Doubtless it was the beautiful plumage of the birds of their forests which first suggested this admirable art: but of these the one held in the greatest respect by them was the humming-bird. It was their belief that Toyamiqui, the spouse of the God of War, conducted the souls of warriors who had died in defence of the gods into the mansion of the sun, and there transformed them into humming-birds; they believed, also, that the humming-bird, like the dove of Noah, went forth from the ark and returned with a twig in its mouth. Thus endeared to them by association, and venerated by tradition, this diminutive bird supplied them with the choicest materials for the art in which they most delighted—the *plumaje* or feather embroidery, with which they could produce all the effects of delicate pictures. The most airy tints of landscape, the most complicated combinations of flowers were alike imitated with marvellous fidelity, and the following anecdote, related by Antonio de Herrara, proves their skill in figure-painting:—" Don Philip, the prince of Spain, his schoolmaster did give unto him three figures, or portraitures, made of feathers, as it were to put in a breviarie. His Highness did show them to King Philip, his father, the which his Majestie beholding attentively said, that he had never seene in so small a worke a thing of so great excellency and perfection. One day as they presented to Pope Sixtus Quintus another square bigger than it, wherein was the figure of St. Francis, and that they had told him it was made of feathers by the Indians, he desired to make a trial thereof, touching the table with his fingers to see if it were of feathers."

We can fancy the worthy old gentleman fingering these beautiful works of art with the curiosity of a schoolboy; but his test was certainly less destructive than that of Peter the Great at Copenhagen, who, being shown a choice mosaic, flattened a pistol bullet against it to decide the fact of its being made of stone! Herrara goes on to say that "they make the best figures of feathers in the province of Mechonacan, and in the village of Poscaro. The manner is, with small delicate pinsors they pull the feathers from the dead fowles, and with a fine paste they cunningly join them together."*

The feathers were, in reality, fixed on a very fine cotton web, and were wrought into dresses for the wealthy, also hangings for palaces and ornaments for the temples. Zuazo extols the beauty and warmth of this fabric, saying—"I saw many mantels worked with feathers of the humming-bird, so soft, that passing the hand over them they appeared to be like hair. I weighed one of these which did not weigh more than six ounces. They say that in the winter one is sufficient over the shirt without any covering, or any other clothes over the bed."

One of the noblest aviaries in the world was that attached to the palace of the ill-fated Montezuma. Here were collected the scarlet cardinal, the golden pheasant, the endless parrot tribe, and hundreds of humming-birds, which delighted to revel in the honeysuckle bowers. Three hundred attendants had charge of this aviary, and in the moulting season it was their especial duty to collect the brilliant plumage for the use of the numerous Sultanas who employed their days in this feather embroidery: old Gomara, who had a fine eye for the picturesque, and who saw

* Antonio de Herrara. Description of the West Indies. Purchas, vol. iii.

the Tlascalian army decked out in all their plumal array, says, "They were trimme felowes and well armed according to their use, although they were painted so that their faces showed like divels, with great tuffes of feathers and triumphed gallantry." Doubtless the scene must have been brilliant, for all the chiefs wore plumes and gorgeously embroidered surcoats, and there were banners and devices worked in gaudy hues, whilst the national standard displayed in exquisite feather-work and gold the armorial ensigns of the state.

> "Others of higher office were arrayed
> In feathery breastplates of more gorgeous hue
> Than the gay plumage of the mountain cock,
> Or pheasant's glittering pride.
>
> . . . .
>
> The golden glitterance, and the feather mail
> More gay than glittering gold; and round the helm
> A coronal of high upstanding plumes
> Green as the spring grass in the sunny shower;
> Or scarlet bright, as in the wintry wood
> The clustered holly; or of purple tint
> Whereto shall that be likened; to what gem
> Indiademed! what flower? what insect's wing?"

Not only was the great hall of justice called the "Tribunal of God," festooned with feather tapestry embroidered in beautiful devices of birds and flowers, but above the throne was a canopy of resplendent plumage, from the centre of which shot forth rays of gold and jewels. But, perhaps, that which consecrated the humming-birds most in the estimation of this superstitious people was its connection with the Mexican God of War. This terrible idol, whose altars constantly reeked with the blood of human sacrifices, was Huitzelopotchli, a name compounded of two words signifying "humming-bird" and "left," from the left foot being decorated with the choicest specimens of this favourite plumage.

Among the presents sent by the ill-fated Montezuma to Cortez, and transmitted by him to the Court of Spain, where from their novelty and beauty they excited the greatest possible sensation, were two birds of featherwork and gold thread, the quills of their wings and tails, their feet, eyes, and beaks being of gold; they stood upon reeds of gold raised on balls of featherwork and gold, with tassels of featherwork hanging from each. There were also sixteen shields of precious stones with brilliant feathers hanging from their rims, five beautiful feather fans, and the choicest specimens of feather-tapestry.

The humming-bird tribe is nearly confined to the tropical portions of the New World; the southern continent as far as the tropic of Capricorn, and the great archipelago of islands between Florida and the mouth of the Orinoco, literally swarm with them. A high temperature is, however, by no means essential for their existence, as the most beautiful species are found at an elevation of from seven to twelve thousand feet above the level of the sea, and one of remarkable brilliancy inhabits Chimborazo, at the height of fifteen thousand feet. Other species live in the dreary climate of Tierra del Fuego; and Captain King saw many of these birds flitting about with perfect satisfaction during a heavy snow-storm near the straits of Magellan. In the humid island of Chiloe, the humming-birds darting between the dripping branches, agreeably enlighten the scene; and Juan Fernandez—sacred to early associations—has two species peculiar to itself. Captain Woodes Rogers, who visited this island in 1708, and took Alexander Selkirk from it, says, "And here are also humming-birds about as big as bees: their bill about the bigness of a pin; their legs proportionable to their body; their feathers mighty small, but of most beautiful colours. They are seldom taken or seen but in the evening, when they fly about, and sometimes when dark into the fire."*

It is from the noise produced by the vibration of its wings that the humming-bird derives its name; for rapidity of flight it is quite without an equal, and to this end the shape and structure of its body beautifully tend. In no birds are the pectoral muscles— the chief agents in flight—so largely developed, and

* Harris's Voyages, vol. i. p. 157.

in none are the wings and the individual feathers so wonderfully adapted for rapid locomotion; the tail, though presenting every conceivable modification of form, is always made available as a powerful rudder, aiding and directing the flight; the feet, too, are singularly and disproportionately small, so that they are no obstruction to its progress through the air. Several species have the feet enveloped in most beautiful fringes of down, as if each were passed through a little muff, either white, red, or black.

The eggs of humming-birds are two in number, white, and of an oblong form; but the nests in which they are contained are almost as marvellous as the birds themselves. What will be said of a nest made of thistledown;—and yet one is to be seen in Mr. Gould's collection. The finest down, the most delicate bark, the softest fungi, the warmest moss—all are made available by the different species of these lovely birds, and not less various are the localities in which the diminutive nests are placed. A tiny object is seen weighing down the streaming leaf of a bamboo overhanging a brook; it is one of these nestlets attached to the point of the fragile support, and waving with it in the breeze. Another tribe prefers the feathery leaves of the fern, whilst the tip of the graceful palm-leaf is the favourite bower of a third species; but in every instance the spot is admirably selected to preclude marauding serpents or monkeys from destroying the eggs and callow young.

The down of the cotton-tree, banded round with threads of spiders' webs, forms the fairy abode of the Mango humming-bird. This silky filamentous down is borne upon the air, and though so impalpable as to be inhaled by man in the breath he inspires, it is diligently collected by these little creatures. They may be seen, suspended in the air, battling with a puff of down, which, sailing with the gentle breeze, coquettishly eludes the stroke of the eager beak; filament after filament is however secured, and borne in triumph to complete the elfin bower.

> "There builds her nest the humming-bird,
> Within the ancient wood,
> Her nest of silky cotton down,
> And rears her tiny brood."

Preparatory to the nidification is the important preliminary of courting, and on this delicate proceeding Mr. Gosse throws light. In a cage were placed two long-tailed males and a female. "The latter interested me much," says he; "for, on the next day after her introduction, I noticed that she had seated herself by a male, on a perch occupied only by them two, and was evidently courting caresses. She would hop sideways along the perch by a series of little quick jumps, till she reached him, when she would gently peck his face and then recede, hopping and shivering her wings, and presently approach again, to perform the same actions. Now and then she would fly over him, and make as if she were about to perch on his back, and practise other little endearments."* We regret to say that the cold-blooded long-tailed gentleman was utterly indifferent to all these delicate attentions, and sat gloomily chewing the cud of his own reflections; a few days afterwards the lady-bird made her escape, and we hope soon ceased to wear the willow.

The same able observer gives the following account of the nest-building of one of these elegant birds. The scene was at a place called Bognie, on the Bluefields Mountain, in Jamaica. "About a quarter of a mile within the woods, a blind path, choked up with bushes, descends suddenly beneath an overhanging rock of limestone, the face of which presents large projections and hanging points incrusted with a rough tuberculous sort of stalactite. At one corner of the bottom there is a cavern, in which a tub is fixed to receive water of great purity, which perpetually drips from the roof, and which, in the dry season, is a most valuable resource. Beyond this, which is very obscure, the eye penetrates to a larger area, deeper still, which receives light from some other communication with

* "The Birds of Jamaica," by P. H. Gosse, 1847.

the air. Round the projections and groins of the front, the roots of the trees above have entwined, and to a fibre of one of these, hanging down, not thicker than a whipcord, was suspended a humming-bird's nest containing two eggs. It seemed to be composed wholly of moss, was thick, and attached to the rootlet by its side. One of the eggs was broken. I did not disturb it, but, after about three weeks, visited it again. It had been apparently handled by some curious child, for both eggs were broken, and the nest was evidently deserted. But while I lingered in the romantic place, picking up some of the land shells which were scattered among the rocks, suddenly I heard the whirr of a humming-bird, and, looking up, saw a female *Polytmus* hovering opposite the nest with a mass of silk cotton in her beak. Deterred by the sight of me, she presently retired to a twig a few paces distant, on which she sat. I immediately sank down among the rocks, as quietly as possible, and remained perfectly still. In a few seconds she came again, and after hovering a moment, disappeared behind one of the projections, whence, in a few seconds, she emerged again and flew off. I then examined the place, and found, to my delight, a new nest—in all respects like the old one—unfinished, affixed to another twig not a yard from it. I again sat down among the stones in front, where I could see the nest, not concealing myself, but remaining motionless, waiting for the *petite* bird's reappearance. I had not to wait long. A loud whirr, and there she was, suspended in the air before her nest. She soon espied me, and came within a foot of my eyes, hovering just in front of my face. I remained still, however, when I heard the whirring of another just above me—perhaps the mate—but I durst not look towards him, lest the turning of my head should frighten the female. In a minute or two the other was gone, and she alighted again upon the twig, where she sat some little time preening her feathers, and apparently clearing her mouth from the cotton fibres, for she now and then swiftly projected the tongue an inch and a half from the beak, continuing the same curve as that of the beak. When she arose, it was to perform a very interesting action; for she flew to the face of the rock, which was thickly clothed with soft dry moss, and, hovering on the wing, as if before a flower, began to pluck the moss, until she had a large bunch of it in her beak. Then I saw her fly to the nest, and, having seated herself in it, proceed to place the new material, pressing and arranging, and interweaving the whole with her beak, while she fashioned the cup-like form of the interior by the pressure of her white breast, moving round and round as she sat. My presence appeared to be no hindrance to her proceedings, though only a few feet distant. At length she left again, and I left the place also. On the 8th of April, I visited the cave again, and found the nest perfected, and containing two eggs, which were not hatched on the 1st of May, on which day I sent Sam to endeavour to secure both dam and nest. He found her sitting, and had no difficulty in capturing her, which, with the nest and its contents, he carefully brought down to me. I transferred it—having broken one egg by accident—to a cage, and put in the bird. She was mopish, however, and quite neglected the nest, as she did also some flowers which I inserted. The next morning she was dead."

The Rev. Lansdown Guilding once saw a pair of humming-birds (*Trochilus cristatus*) domesticated in the house of a gentleman. They built for many years on the chain of the lamp suspended over the dinner-table, and here they educated several broods in a room occupied hourly by the family. He has been seated with a large party at the table when the parent bird has entered, and passing along the faces of the visitors, displaying his gorgeous crest, has ascended to the young without fear or molestation. The same writer has seen a bird of this species caught and nearly perishing in the nets of a large spider, from which it was unable to extricate itself.

When looking at humming-birds—some not bigger

than a humble-bee, and blazing with all the refulgence of the brightest jewels—it is scarcely possible to imagine how they can be obtained without serious damage to their beauty. Some writers have stated that they are shot with charges of sand; others, that water is the missile—but they are mistaken; various methods are certainly employed, but neither of those. The little creatures are sometimes shot with small charges of "dust-shot," as the smallest pellets are called; frequently the keen eye and steady hand of the Indians bring them down by an arrow from their blow-tube; a third mode is to watch them into a deep tubular flower, and to secure them with a gauze net which is skilfully thrown over it.

Very many humming-birds were caught by Mr. Gosse, with a common gauze butterfly-net, on a ring a foot in diameter. The curiosity of humming-birds is great; and on holding up the net near one, he frequently would not fly away, but come and hover over the mouth, stretching out his little neck to peep in. Often, too, when an unsuccessful stroke was made, the bird would return immediately, and suspend itself in the air, just over his pursuer's head, or peep into his face with unconquerable familiarity. But, when caught, they usually soon died; they would suddenly fall to the floor of the cage, and lie motionless, with closed eyes. If taken into the hand, they would perhaps seem to revive for a few moments, then throw back the pretty head, or toss it to and fro as if in great suffering, expand the wings, open the eyes, slightly puff the feathers of the breast, and die. Such was the result of his first efforts to procure these birds alive; but he was subsequently more fortunate.

Collecting the nests of humming-birds in the West Indies, requires some care, on account of the great number of venomous serpents which frequent the thickets.

While Alexander Wilson, the subsequently celebrated ornithologist, was struggling against poverty in his early days as a weaver, he was much importuned by a shopmate to write him an epitaph. This individual had excelled in little, except, to use the expressive Scottish word, *daundering* about the hedgerows on Sundays, in search of birds' nests. After much pressing, Wilson complied, and hit off the following:—

> "Below this stane John Allan rests,
> An honest soul, though plain;
> He sought hail Sabbath day for nests,
> But always sought in vain."

Had Mr. Allan pursued his nidal investigations in Jamaica, his curiosity might have met with an unpleasant check. A young gentleman of similar tastes, observing a parroquet enter a hole in a large duckant's nest situated on a bastard cedar, mounted to take her eggs or young. Arrived at the place, he cautiously inserted his hand, which presently came into contact with something smooth and soft; he thought it might be the callow young, but having some misgivings, descended and procured a stick; having again mounted, he thrust in the stick, and forced off the whole upper part of the structure, when to his utter discomfiture and terror, an enormous yellow boa was disclosed, his jaws retaining the feathers of the parroquet, which had just been swallowed. The serpent instantly darted down the tree, and the curious youth descended scarcely less rapidly, and fled, cured for a time of bird-nesting.

A story is told of a trick played upon an enthusiastic foreign naturalist, on his landing at Rio Janeiro, by certain middies of the ship which had carried him out. The worthy *savant* was very stout, very nearsighted, and very eager to collect humming-birds. The young gentlemen therefore determined to make merry at his expense in the following manner:— Having caught several large blue-bottle flies, they stuck them over with small bits of gay peacock feathers, with two long plumules behind, by way of tail; the wings were left free. Then carefully placing the chairs, boxes, and crockery of the doctor's apartment in every possible direction, they turned their

insect "daws" loose into the room, and quietly
awaited the result in the adjoining chamber. Presently the victim was heard creaking slowly up the
stairs, anathematizing the heat and puffing for breath.
He entered his room, the door closed, and there was
a pause. Very shortly, a tremendous scuffling and
rushing about commenced; chairs were heard to fall,
crockery to break, and at last the smash of a looking-glass completed the scene. The wags now entered
the room, and found the doctor with his coat off in a
state of great excitement; his eyes were filled with
tears, and he was actively rubbing one of his shins.
"Good gracious! my dear sir, what's the matter?
Is it a *coup de soleil*, or—the brandy, eh!" "No,
sare; neither one nor de oder," replied he, with
intense earnestness; "I was catch de charmant
littel bottel-blue homing bairds, but dey be so *dam*
wild." His indignation, when the explosion of now
irrepressible laughter proclaimed the trick, was marvellous to behold.

Wilson, in his 'American Ornithology,' states,
that Mr. C. W. Peale told him that he had two
young humming-birds, which he had raised from the
nest. They used to fly about the room, and would
frequently perch on Mrs. Peale's shoulder to be fed.
When the sun shone strongly in the chamber, they
have been seen darting after the motes that floated

in the light, as fly-catchers would after flies. In the summer of 1833, a nest of young humming-birds, nearly ready to fly, was brought to Wilson himself. One of them flew out of the window the same evening, and falling against a wall, was killed; the other refused food, and the next morning was all but dead; a lady undertook to be the nurse of this lonely one, placed it in her bosom, and as it began to revive, dissolved a little sugar in her mouth, into which she thrust its bill, and it sucked with great avidity; in this manner it was brought up until fit for the cage. Mr. Wilson kept it upwards of three months, supplied it with loaf-sugar dissolved in water, which it preferred to honey and water, and gave it fresh flowers every morning sprinkled with the liquid. It appeared gay, active, and full of spirit, hovering from flower to flower, as if in its native wilds; and always expressed, by its motions and chirping, great pleasure at seeing fresh flowers introduced to its cage; every precaution was supposed to have been taken to prevent its getting at large, and to preserve it through the winter; but unfortunately it by some means got out of its cage, and, flying about the room, so injured itself that it soon died. A striking instance is mentioned by the same author, of the susceptibility of some humming-birds to cold; in 1809, a very beautiful male was brought to him, put into a wire cage, and placed in a shady part of the room, the weather being unusually cold; after fluttering about for some time, it clung by the wires, and hung in a seemingly torpid state for a whole forenoon; no motion of respiration could be perceived, though at other times this is remarkably perceptible; the eyes were shut, and when touched by the fingers the bird gave no signs of life or motion; it was carried into the open air, and placed directly in the rays of the sun, in a sheltered situation. In a few seconds respiration became apparent; the bird breathed faster and faster, opened its eyes, and began to look about with as much vivacity as ever. After it had completely recovered it was restored to liberty, and flew off to the withered top of a pear-tree, where it sat for some time, dressing its disordered plumage, and then shot off like a meteor.

Though some humming-birds are gifted with powers of song, the greater number give utterance to a note not unlike the scraping of two boughs, one against the other. The following spirited description of Mr. Nutall, of that beautiful species, the ruff-necked humming-bird, applies very generally to the class. "We now for the first time saw the males in numbers, darting, burring, and squeaking in the usual manner of their tribe; but when engaged in collecting sweets in all the energy of life, it seemed like a breathing gem or magic carbuncle of glowing fire, stretching out its gorgeous ruff as if to emulate the sun itself in splendour. Towards the close of May, the females were sitting, at which time the males were uncommonly quarrelsome and vigilant, darting out at me as I approached the tree, probably near the nest, looking like an angry coal of brilliant fire, passing within very little of my face, returning several times to the attack, striking and darting with the utmost velocity, at the same time uttering a curious, reverberating, sharp bleat, somewhat similar to the quivering twang of a dead twig, yet also so much like the real bleat of some small quadruped, that for some time I searched the ground instead of the air for the actor in the scene. At other times the males were seen darting up high in the air, and whirling about each other in great anger and with much velocity."

The luxuriance of tropical vegetation, and the varied richness of its hues, is dwelt upon with admiration by all travellers; and such a scene as the following is a fit palace for Nature's most glorious gems; it is a description of 'the Bluefields Ridge' in Jamaica. "Behinds the peaks, which are visible from the sea at an elevation of about half a mile, there runs through the dense woods a narrow path just passable for a horse, overrun with beautiful ferns of many graceful forms, and always damp and cool. . . . The refreshing coolness of this road, its unbroken solitude, combined with the peculiarity and luxuriance of its vegetation, made it one of my favourite resorts. Not a tree, from the thickness of one's wrist up to the giant magnitude of the hoary figs and cotton-trees, but is clothed with fantastic parasites; begonias with waxen flowers, and ferns with hirsute stems, climb up the trunks; enormous bromelias spring from the greater forks, and fringe the horizontal limbs; various orchideæ, with matted roots and grotesque blossoms, droop from every bough, and long lianas, like the cordage of a ship, depend from the loftiest branches, or stretch from tree to tree. Elegant tree-ferns and towering palms are numerous. Here and there the wild plantain or heliconia waves its long flag-like leaves from amidst the humbler bushes, and, in the most obscure corners over some decaying log, nods the noble spike of a magnificent limodorum. Nothing is flaunting or showy: all is solemn and subdued; but all is exquisitely beautiful. Now and then the ear is startled by the long-drawn measured notes, mostly richly sweet, of the *solitaire*, itself mysteriously unseen, like the hymn of praise of an angel." Such is the glorious scene rendered still more attractive by the long-tailed humming-bird, which resorts thither in hundreds to feed on the scarlet "glass-eye berries." These little fellows, and indeed the whole tribe, are so pugnacious, that two of the same species can rarely suck flowers from the same bush without a rencontre. The Mango exceeds all others in pugnacity, and Mr. Gosse describes a scene of which he was witness. "In the garden were two trees of the tribe called the Malay apple, one of which was but a yard or two from my window. The genial influence of the spring rains had covered them with a profusion of beautiful blossoms, each consisting of a multitude of crimson stamens, with very minute petals, like bunches of crimson tassels. A mango humming-bird had every day, and all day long, been paying his devoirs to these charming blossoms. On the morning to which I allude, another came, and the manœuvres of these two tiny creatures became highly interesting. They chased each other through the labyrinth of twigs and flowers, till an opportunity occurring, the one would dart with seeming fury upon the other, and then with a loud rustling of their wings, they would twirl together, round and round, until they nearly came to the earth. It was some time before I could see with any distinctness what took place in these tussles: their twirlings were so rapid as to baffle all attempts at discrimination. At length an encounter took place pretty close to me, and I per-

ceived that the beak of the one grasped the beak of
the other, and thus fastened both twirled round and
round in their perpendicular descent, the point of
contact being the centre of the gyrations, till, when
another second would have brought them both on the
ground, they separated, and the one chased the other
for about a hundred yards, and then returned in
triumph to the tree, where, perched on a lofty twig,
he chirped monotonously and pertinaciously for some
time;—I could not help thinking, in defiance; in a
few minutes, however, the banished one returned,
and began chirping no less provokingly, which soon
brought on another chase and another tussle. . . .
Sometimes they would suspend hostilities to suck a
few blossoms, but mutual proximity was sure to
bring them on again with the same result. In their
tortuous and rapid evolutions the light from their
ruby necks would now and then flash in the sun with
gem-like radiance, and as they now and then hovered
motionless, the broadly expanded tail—whose outer
feathers are crimson purple, but when intercepting
the sun's rays transmit orange-coloured light—added
much to their beauty. A little Banana Quit, that
was peeping among the blossoms in his own quiet
way, seemed now and then to look with surprise on
the combatants; but when the one had driven his
rival to a longer distance than usual, the victor set

upon the unoffending Quit, who soon yielded the point, and retired humbly enough to a neighbouring tree."

The flight of the humming-bird from flower to flower has been described as resembling that of the bee, but so much more rapid, that the latter appears a loiterer by comparison. The bird poises himself on wing while he thrusts his long slender tubular tongue into the flowers in search of honey or of insects; he will dart into a room through an open window, examine a bouquet of flowers with the eye of a connoisseur, and, *presto!* is gone. One of these birds has been known to take refuge in a hot-house during the cold autumnal nights, leaving it in the morning and returning regularly every evening to the chosen twig in its warm palace.

> " For though he hath countless airy homes
> To which his wing excursive roves;
> Yet still from time to time he loves
> To light upon earth, and find such cheer
> As brightens his banquet here."

The Mango Humming-Bird is familiarly known to the negroes of Jamaica by the name of the "Doctor-Bird," said to have been thus derived. In the olden time, when costume was more observed than now, the black livery of this bird among its more brilliant companions, bore the same relation as the sombre costume of the grave physician to the gay colours then worn by the wealthy planters, whence the humorous comparison and name. It might, with equal propriety, have been called the Parson, but in those days ecclesiastics were but little known by the negroes.

Mr. Gosse observed that the bunch of blossoms at the summit of the pole-like papaw-tree is a favourite resort of this species, and, taking advantage of this, succeeded in catching a fine live specimen. " Wishing," says he, " to keep these birds in captivity, I watched at the tree one evening with a gauze ring-net in my hand, with which I dashed at one, and though I missed my aim, the attempt so astonished it, that it appeared to have lost its presence of mind, so to speak, flitting hurriedly hither and thither for several seconds before it flew away. The next evening, however, I was more successful. I took my station and remained quite still, the net being held up close to an inviting bunch of blossoms; the humming-birds came near in their course round the tree, sipped the surrounding blossoms, eyeing the net; hung in the air for a moment in front of the fatal cluster, without touching it, and then, arrow-like, darted away. At length one, after surveying the net, passed again round the tree; on approaching it the second time, perceiving the strange object to be still unmoved, he took courage and began to suck. I quite trembled with hope; in an instant the net was struck, and, before I could see anything, the rustling of his confined wings within the gauze told that the little beauty was a captive. I brought him in triumph to the house and caged him, but he was very restless, clinging to the sides and wires and fluttering violently about. The next morning, having gone out on an excursion for a few hours, I found the poor bird on my return dying, having beaten himself to death.

Two young males of the long-tailed species were subsequently captured, and, instead of being caged, they were turned loose into a room. They were lively, but not wild; playful towards each other, and tame to their captor—sitting on his finger, unrestrained, for several seconds at a time: on a large bunch of asclepias being brought into the room, they flew to the nosegay and sucked while in Mr. Gosse's hand: these and other flowers being placed in glasses, they visited each bouquet in turn, sometimes playfully chasing each other, and alighting on various objects. As they flew they were repeatedly heard to snap the beak, at which time they doubtless caught minute flies. After some time one of them suddenly sank down in one corner, and on being taken up, seemed dying; it had perhaps struck itself during its flight: it lingered awhile and died.

Another of these long-tailed humming-birds brought alive to Mr. Gosse, became so familiar that even before he had had the bird a day it flew to his face, and perching on his lip or chin, thrust its beak into his mouth. It grew so bold and so frequent in its visits as to become almost annoying, thrusting its protruded tongue into all parts of his mouth in the most inquisitive manner. Occasionally his master gratified it by taking a little syrup into his mouth, and inviting him to the banquet by a slight sound, which it soon learned to understand. Mr. Gosse had now several pets of this beautiful species, and it was interesting to observe how each selected its own place for perching and for roosting, to which it invariably adhered, a peculiarity which caused many others to be caught; for, by observing a place of resort, and putting a little birdlime on that twig, a bird would be captured in a few minutes.

Of the birds in this gentleman's possession, one would occasionally attack a gentler and more confiding companion, who always yielded and fled, whereupon the little bully would perch and utter a cry of triumph in a succession of shrill chirps. After a day or two, however, the persecuted one would pluck up courage, and play the tyrant in turn, interdicting his fellow from sipping at the sweetened cup. Twenty times, in succession, would the thirsty bird drop down on the wing to the glass, but no sooner was he poised and about to insert his tongue than the other would dart down, with inconceivable swiftness, and, wheeling so as to come up beneath him, would drive him away from the repast: he might fly to any other part of the room unmolested, but an approach to the cup was a signal for an instant assault. When these birds had become accustomed to the room, their vivacity was extreme, and their quick turns caused their lovely breasts to flush out from darkness into sudden lustrous light, like rich gems. Their movements in the air were so rapid as to baffle the eye. Suddenly the radiant little meteor would be lost in one corner, and as quickly the vibration of its invisible wings would be heard behind the spectator; in another instant it would be hovering in front of his face, curiously peering into his eyes with its own bright little orbs.

Of twenty-five of the species taken, only seven were domesticated, and there was much difference in the tempers of these; some being moody and sulky, others wild and timid, and others gentle and confiding from the first.

It is just possible that these pages may be perused by some one under favourable circumstances for the capture of humming-birds. To them the following remarks, founded on the experience of Mr. Gosse, may prove acceptable. There should be a very capacious cage, wired on every side, in the bottom of which a supply of decaying fruit, as oranges or pines, should be constantly kept, but covered with wire, that the birds may not soil their plumage. This would attract immense numbers of small flies, which would, in conjunction with syrup, afford food for the birds. It was observed that on opening the basket in which newly caught humming-birds were confined, they would fly out and soar to the ceiling, rarely seeking the window. There they would remain on rapidly vibrating pinions, lightly touching the plaster with their beak or breast every second, and slightly rebounding: after a time they became exhausted, and sank rapidly down to alight; they would then suffer themselves to be raised, applying their little feet to a finger passed under the breast; they were then gently raised to their captor's mouth, and would generally suck syrup from the lips with eagerness. When once fed from the mouth the birds were always ready to suck afterwards, and frequently voluntarily sought the lips: after a time a glass of syrup was presented to the bird instead of the lips, and it soon learned to sip from this, finding it as it stood on a table; it was then considered domesticated.

Not the least curious part of the structure of humming-birds is the tongue, which consists of two tubes laid side by side like a double-barrelled gun, but separated at a short distance from the tip, where each is somewhat flattened. This tongue is connected with a very beautiful apparatus, whereby it can be darted out to a great length, and suddenly retracted. The food of humming-birds consists of insects and the honeyed juices of flowers, and with this tongue the latter are pumped up. The mode of catching insects is interesting. "I have," says Mr. Bullock,* "frequently watched, with much amusement, the cautious peregrination of the humming-bird, who advancing beneath the web (of the spiders) entered the various labyrinths and cells in search of entangled flies; but as the larger spiders did not tamely surrender their booty, the invader was often compelled to retreat: being within a few feet I could observe all their evolutions with great precision. The active little bird generally passed once or twice round the court as if to reconnoitre his ground, and commenced his attack by going carefully under the nets of the wily insect, and seizing by surprise the smallest entangled insects, or those that were most feeble. In ascending the angular traps of the spider, great care and skill were required; sometimes he had scarcely room for his little wings to perform their office, and the least deviation would have entangled him in the complex machinery of the web, and involved him in ruin. It was only the works of the smaller spiders that he durst attack, as the largest rose to the defence of their citadels, when the be-

* Six Months in Mexico.

sieger would shoot off like a sunbeam, and could only be traced by the luminous glow of his refulgent colours. The bird generally spent about ten minutes in this predatory excursion, and then alighted on a branch of an avocata to rest and refresh himself, placing his crimson starlike breast to the sun, and when there presented all the glowing fire of the ruby, and surpassed in lustre the diadem of a monarch."

The mode in which the humming-birds in Mr. Gosse's possession partook of their favourite banquet of syrup was very characteristic. These volatile geniuses would not condescend to such a matter-of-fact proceeding as to fly straight to the glass—by no means; they invariably made a dozen or twenty distinct stages or swoops, each in a curve descending a little, then ascending again and hovering a second or two at each angle. Sometimes when they arrived opposite the cup more quickly than was intended, they would retreat again, as if, as with hydropathic patients, a certain number of " turns" were indispensable before breakfast. When this proceeding was completed, and the appetite had acquired the proper razor-edge, they would bring their tiny feet to the edge of the glass, insert the sucking tongue in its contents, and take a draught of nectar.

One of the earliest notices of humming-birds occurs in the writings of Antonio de Herrera, who, by the way, rejoiced in the title of "*Coronesta Mayor de las Indias y Castilla,*" and died in 1625. In his "Historia General" he says, "There are some birds in the country (Mexico) of the size of butterflies, with long beaks and brilliant plumage, much esteemed for the curious works made of them. Like the bees they live on flowers and the dew which settles on them; and when the rainy season is over, and the dry weather sets in, they fasten themselves to the trees by their beaks, and soon die; but in the following year, when the new rains come, they come to life again." The same writer says, that the women and girls of the Caribbee Islands, especially Martinico, hung humming-birds from their ears as pendants, and very elegant ornaments they doubtless were. Our knowledge of the tribe is very recent. Linnæus was not acquainted with more than half-a-dozen species; and Mr. Bullock states, in the catalogue of his well-known Museum, in 1812, "This case contains nearly one hundred various humming-birds, and is allowed to be the finest collection in Europe." Subsequently, the late Mr. George Loddiges, of Hackney, brought together a very noble collection; but it was reserved for Mr. Gould, the great historian of the feathered race, to display to their full extent the marvellous treasures of this wonderful tribe. In his collection, at the present moment (May, 1851), one of the great features of the Zoological Gardens, Regent's Park, there are upwards of three hundred species, and more than two thousand birds. Within the last two months, a wooden building, in the Swiss style, has been erected for its reception on the south side of the gardens, the dimensions being sixty feet by thirty. It is lighted exclusively from above, the glare being mellowed by neat calico hangings, suspended beneath the skylights. The walls are papered with very elegant and appropriate designs, and are adorned with framed illustrations from Mr. Gould's work on the "Trochilidæ,"—themselves marvels of art; the general effect of the apartment is most charming. The birds themselves are in glass cases, mounted in white and gold, and disposed in three rows down the room. Consummate skill has been displayed in the arrangement of their contents, in order that the colours or peculiarities of each group may be displayed to the greatest advantage. Here, for instance, are a pair of exquisite little creatures, absorbed in amorous dallying: there, two blazing gems, engaged in furious combat. In one spot, birds are tranquilly pluming their feathers, and in another they are sipping imaginary nectar from choice flowers. Not only are the birds shown, but their habits are still further elucidated by the introduction of their nests and eggs; indeed, the illusion of reality is complete, from the skill with which some are displayed feeding the half-fledged young, whilst other proud mothers are sitting in their Lilliputian nests, apparently as in life, listening to the tender lays of their more gorgeous mates.*

Not the least remarkable feature of this collection, is the short period in which it has been formed; though to those acquainted with the energetic character of Mr. Gould, it is scarcely matter of surprise. That gentleman turned his attention for the first time to the subject, on the conclusion of his great work on the birds and mammals of Australia, about five years ago; entering on it with a zeal for which he is unsurpassed, he soon found himself in possession of sufficient materials to publish the first fasciculus of a work on 'Trochilidæ,' or Humming-Birds, which bids fair to be absolutely unrivalled. Shortly after this had been issued, the author discovered a method of painting on gold and silver, whereby he succeeded in attaining the long-sought desideratum, that beautiful metallic flashing lustre so marked in this class. Regardless of the heavy sacrifice, he immediately recalled the copies which had been issued, and replaced them with others, coloured in the improved manner. It is with specimens of these plates that the walls of the room are adorned.

* This unique exhibition was visited by 80,000 persons during the nine months it was open.

The sight of this collection can scarcely fail to excite a desire to possess some living specimens of these precious gems. It may be done with less difficulty than is imagined, by attending to the instructions already given, and by caging the birds immediately before the sailing of the vessel. If forwarded by a steamer, large bunches of fresh flowers might be obtained for their refreshment at St. Thomas, Bermuda, and the Azores; and thus we may hope to see these lovely creatures domesticated in our conservatories, as we have already seen far rarer and more singular animals acclimated in our menageries.

CHAPTER IX.

CROCODILES.—ANCIENT WRITERS.—MODE OF CAPTURE.—SACRED CROCODILES.—TENTYRITES.—RARE BOOK.—INDIAN WORSHIP—MEDICINAL VIRTUES.—CROCODILES AND ALLIGATORS.—ANATOMICAL PECULIARITIES. —TEETH. — NIDIFICATION. —CROCODILES AND TROCHILUS.—THE ZIC-ZAC.—CROCODILE BIRD. - HYBERNATION.—JACARÉS OF THE AMAZON.—POACHERS.—MR. SPRUCE.—ANECDOTES.—SEARCH FOR VICTORIA REGIA.—A DISAGREEABLE NEIGHBOUR—THE BATTLE.—THE DEATH.—A DAINTY LUNCHEON. —ALLIGATORS AND DOGS.—MR. WATERTON.—RIDING ON CROCODILES.—A BOLD AFRICAN.—ALLIGATOR TANK.—THE SUBALTERN'S SPORT.—CONCLUSION.

IT might have been anticipated that an animal which abounded in the great river of Egypt in the time of the Israelites, and was an object of idolatrous worship to the inhabitants, should have attracted the notice of the inspired writers of old: accordingly, various allusions to it are found in the sacred writings. Commentators, however, differ as to whether it is the crocodile which, under the name of Leviathan, forms the subject of one of the sublimest chapters of Job; the description is applicable in some respects, and not in others; but there can be little doubt that this creature is referred to under the Hebrew name *Than*, translated *dragon*, in the following figurative passage of Ezekiel.

"Behold, I am against thee, Pharaoh king of Egypt, the great dragon that lieth in the midst of his rivers, which hath said, My river is mine own, and I have made it for myself. But I will put hooks in thy jaws, and I will cause the fish in thy rivers to stick unto thy scales; and I will bring thee up out of the midst of thy rivers, and all the fish of thy rivers shall stick unto thy scales."*

Here we have a distinct allusion to the mode of taking the crocodile practised by the Egyptians, as described by Herodotus, whose statements upon these and some other disputed points, have been proved to be trustworthy.

"The modes of taking the crocodile are many and various, but I shall only describe that which seems to me most worthy of relation. When the fisherman has baited a hook with the chine of a pig, he lets it down into the middle of the river, and holding a young live pig on the brink of the river, beats it. The crocodile hearing the noise, goes in its direction, and meeting with the chine, swallows it; but the men draw it to land. When it is drawn out on shore, the sportsman first of all plasters its eyes with mud; and having

* Ezekiel xxix. 3, 4.

done this, afterwards manages it very easily; but until he has done this he has a great deal of trouble."*

With the ancient Egyptians, the crocodile was typical of the sun, and Savak, the crocodile-headed deity of Ombos, was a deified form of the sun. In Lower Egypt it was held in especial veneration at a place called the City of Crocodiles, afterwards Arsinoe, and the animals were there kept in the lake Mœris, and when dead, were buried in the underground chambers of the famous Labyrinth.

These sacred crocodiles led a most luxurious life; they were fed with geese, fish, and various delicacies dressed purposely for them; their heads were adorned with ear-rings, their throats with necklaces of gold or artificial stones, and their feet with bracelets. Strabo gives a curious account of an interview he had with one of these portly reptilians. His host was a man of consideration, and anxious to do the honours of the place with becoming courtesy. Having therefore entertained the great geographer and his party at an elegant *déjeûner*, he proposed that they should pay their respects to his neighbour "Souchos." Providing himself with a cake, a loaf of bread, and a cup of wine, he led the way to the borders of the lake, where his crocodilian highness lay stretched in pampered indolence. To open its own mouth was too

* Herodotus: Euterpe, ii. 70.

much trouble, so one of the priests politely did it for him; another put in first the cake, then the meat, which it gratified them by swallowing, and then pledged them in the cup of wine, which was poured down its throat. Having rested awhile after this exertion, his highness entered the lake, crossed it, and submitted to a similar ordeal on the other side, for the gratification of another party who had come to offer their tribute of good things.

Happy were these sacred crocodiles during life, and after death they were not less well cared for;—their bodies were embalmed in a sumptuous manner, and deposited in catacombs hewn out of the limestone rock. There are many of these mummies in the British Museum, all having the same character, that of being rolled and swathed up in oblong packages, carefully and neatly secured with bandages.

It was not, however, throughout the whole of Egypt that this golden age of crocodiles reigned; an iron age overshadowed the race in the land of Tentyris. By its inhabitants they were held in abhorrence, and no opportunity of destroying them was lost; indeed these Tentyrites are said to have been so expert in their pursuit, that they thought nothing of following a crocodile into the Nile and bringing it by force to shore. The following is Pliny's account of this proceeding:—

"The men are but small of stature, but in this quarrell against the crocodiles, they have hearts of lions, and it is wondrous to see how resolute and courageous they are in this behalfe. Indeed this crocodile is a terrible beast to them that flie from him; but contrary, let men pursue him, or make head againe, he runs away most cowardly. Now these islanders be the only men that dare encountre him affront. Over and besides, they will take the river and swim after them; nay, they will mount upon their backs, and set them like horsemen; and as they turne their heads with their mouth wide open, to bite or devour them, they will thrust a club or great cudgell into it, crosse overthwart, and so holding hard with both hands each end thereof, the one with the right, the other with the left, and ruling them perforce, as it were, with a bit and bridle, bring them to land like prisoners. When they have them there, they will so fright them only with their words and speech that they compel them to cast up and vomit those bodies again to be enterred, which they had swallowed but newly before."

There is a very rare and curious book on field sports, by one Johannes Stradaen, in which men are represented riding on crocodiles, and bringing them to land, whilst others are being killed with clubs. The sketch is full of spirit, and below it are these lines:—

Tentyra in Ægypto, Nilum juxtâ insula gentem
Intrepidam gignit: crocodili hæc scandere dorsum
Audet: refrenat baculo os; discedere cogit
Ex amne in terram : mortem acceleratque nocenti."

Strabo bears testimony to the dexterity of the Tentyrites, stating, that when some crocodiles were exhibited in the Circus at Rome, in a huge tank of water, a party of Tentyrites who had accompanied them, boldly entered the tank, and entangling the crocodiles in nets, dragged them to the bank and back again into the water.

Singular to say, homage to these reptiles is still paid in certain parts of India; and the following account, by an eye-witness, almost carries us back to the time of the Egyptians:—

"Among the outlying hills that skirt the Hala Mountains, about nine miles from that town (Karáchí), is a hot spring, the temperature of which where it wells from the earth is 136° of Fahrenheit. The stream irrigates a small valley, and supplies some swamps with water, in which the fakirs keep numbers of tame alligators. The pond where we saw the congregated herd at feeding-time was about eighty yards long, and perhaps half as many wide. It was shallow, and covered with small grass-covered islets, the narrow channels between which would only admit of a single alligator passing through at a time. Two goats were slaughtered for that morning's repast, during which operation a dozen scaly monsters rose out of their slimy bed, crawled up the back of the tank, and eyed, with evident satisfaction, the feast preparing for them. All being ready, a little urchin, not nine years old, stepped without our circle, and calling "*Ow! ow!*" (come, come,) the whole tribe was in motion; and as soon as the amphibious animals had gained *terra firma*, the meat was distributed. Each anxious to secure a piece at his neighbour's expense, the scene that ensued was ludicrous enough, and not a little disgusting. A hind quarter of a goat gave rise to a general engagement. One by one the combatants drew off, till the prize remained in the grasp of two huge monsters. Their noses all but touching, each did his best to drag the bloody morsel from the jaws of his adversary, and a long struggle ensued, in which, by turning and tossing, writhing and twisting, they strove for the mastery. It was a drawn battle, for the leg was torn asunder, and each retained his mouthful, when, with heads erect, they sought the water, showing, as they crawled along, considerable tact in avoiding their less successful neighbours." *

According to Pliny, much medicinal virtue rested in defunct crocodiles. "The blood," he tells us, "mundifieth the eies;" the fat is an excellent depi-

* A Personal Narrative of a Journey to the Source of the River Oxus. By Lieut. John Wood.

latory, and in the language of quaint old Holland, "No sooner is the hare rubbed therewith, but presently it sheddeth." The choicest morsel, however, was "the crocodile's heart wrapped within a lock of wooll which grew upon a black sheepe, and hath no other color medled therewith, so that the said sheepe were the first lambe that the dam yeaned." And this dainty bit answered the same end as quinine with us, driving away quartan agues.

The true *crocodiles* are found in the Old and New World, and especially abound in Asia and Africa. The *alligators* are peculiar to America, and the *gavials* appear to be limited to the Ganges, and other large rivers of continental India; but of all countries America abounds most in these scourges of the river, possessing no less than five species of alligators, and two of crocodiles.

Water is the natural element of the class, and to it they hasten at the least alarm; on land they are encumbered by their heavy tails, which, however, may be used as powerful weapons of offence, for, like the shark, the crocodile can strike a tremendous blow with his tail. The reason that these creatures are unable to turn quickly on land is, that on each side of the vertebræ of the neck there is attached a sort of rib, and the extremities of these ribs meeting along the whole neck, the animal is prevented turning its head to either side, and its movements generally are stiff and constrained.

The general characters of the crocodiles and alligators are, long flat heads, with extremely large mouths extending considerably behind the eyes, thick necks, and bodies protected by regular transverse rows of square bony plates, elevated in the centre into keel-shaped ridges, and disposed on the back of the neck into groups of different forms and numbers according to the species. The tongue is short, and so completely attached to the lower jaw as to be quite incapable of protrusion, hence the ancients believed that the crocodile had no tongue. "This beast alone, of all other that keep the land, hath no use of a tongue," says Pliny. At the back of the mouth there is a structure having special reference to the circumstances under which they usually take their prey; it consists of a valvular apparatus, which cuts off all communication with the throat so effectually, that not a drop of water can enter it, though the mouth be wide open under water. The nostrils are at the tip of the snout, and open into the throat behind the valve. The jaws are also so formed that the nose can be lifted up; by these provisions, the crocodile is enabled to leisurely drown its prey by holding it down, whilst its own breathing is carried on through its nostrils, just elevated above the surface of the water.

Professor Owen has pointed out how admirably the structure and development of Crocodilians are adapted to their nature and habits; and it is interesting to find proof in the fossil jaws of extinct crocodiles which swarmed on the globe countless ages ago, that the same laws regulated the growth and succession of their teeth, as are in force in their existing representatives. Crocodiles come into the world fully equipped with weapons of offence and defence; the number of teeth is as great in the little wriggling wretch just emerged from the egg, as in the patriarchal monster of thirty feet, and it thus arises: the conical sharp-pointed teeth are set in the jaw in a single row, the base of each tooth being hollow; into this cavity the germ of a new and larger tooth fits, and as it grows it reduces the fang of the former by absorption, until, losing all hold, it is pushed out, the new tooth taking its place. This shedding of teeth is in progress during the whole life of the animal.

Herodotus remarks that the crocodile, " of all living beings, from the least beginning, grows to be the largest, for it lays eggs little larger than those of a goose, and the young is at first in proportion to the egg, but when grown up, it reaches to the length of seventeen cubits and even more." This statement of the ancient historian is correct, for the female lays from fifty to sixty eggs, not much larger than those of a goose. She then buries them in the sand, to be hatched by the heat of the sun: and, says Mr. Hill, "just as the period of hatching is completed, exhibits her eagerness for her offspring in the anxiety with which she comes and goes, walks round the nest of her hopes, scratches the fractured shell, and by sounds which resemble the bark of a dog, excites the half-extricated young to struggle forth into life. When she has beheld, with a mixture of joy, fear, and anxiety, the last of her offspring clear of its broken casement, she leads them forth into the pools away from the river, to avoid the predatory visits of the father, who ravenously seeks to prey upon his own offspring." The researches of palæontologists have discovered an interesting fact, that the Plesiosaurus, an early inhabitant of this planet, had a similar propensity, for the bones of young Plesiosauri are found in the petrified excrement of the old ones.

Mr. Edwards, in his interesting voyage up the Amazon, gives valuable information from observation respecting alligators and their nidification. "Soon after," says the writer, "we arrived at the spot which we had marked in the morning where an alligator had made her nest, and *sans cérémonie* proceeded to rifle it of its riches. The nest was a pile of leaves and rubbish nearly three feet in height, and about four in diameter, resembling a cock of hay. We could not imagine how or where the animal had collected such a heap, but so it was. And deep down, very near the surface of the ground, from an even bed, came forth egg after egg, until forty-five had tolerably filled our basket.... These eggs are four inches in length, and oblong, being covered with a crust rather than a shell. They are eaten, and our friends at the house would have persuaded us to test the virtues of an alligator omelette, but we respectfully declined, deeming our reputation sufficiently secured by a breakfast on the beast itself."

There is another point relative to the natural history of the crocodile, mentioned by Herodotus, which has given rise to most conflicting opinions; it is the following:—"All other birds and beasts avoid him, but he is at peace with the trochilus, because he receives benefit from that bird. For, when the crocodile gets out of the water on land, and then opens its jaws, which it does most commonly towards the west, the trochilus enters its mouth, and swallows the leeches which infest it. The crocodile is so well pleased with this service that it never hurts the trochilus." This statement has been regarded by the majority of writers, including Sir Gardner Wilkinson, as a mere myth. Geoffroy St. Hilaire, however, investigated the subject with care, and arrived at the conclusion that there was good foundation for the story of the ancient writer. Mr. Curzon, the author of that delightful work, 'A Visit to the Monasteries of the Levant,' adds his testimony, which is valuable, as coming from a perfectly unprejudiced source.

"I will relate a fact in natural history which I was fortunate enough to witness, and which, although it was mentioned so long ago as the times of Herodotus, has not, I believe, been often observed since; indeed I have never met with any traveller who has himself seen such an occurrence. I had always a strong predilection for crocodile-shooting, and had destroyed several of these dragons of the waters. On one occasion I saw, a long way off, a large one, twelve or fifteen feet long, lying asleep under a perpendicular bank, about ten feet high, on the margin of the river. I stopped the boat at some distance, and, noting the place as well as I could, I took a circuit inland, and came down cautiously to the top of the bank, whence, with a heavy rifle, I made sure of my ugly game. I had already cut off his head in my imagination, and was considering whether it should be stuffed with its mouth open or shut. I peeped over the bank; there he was, within ten feet of the sight of the rifle. I was on the point of firing at his eye, when I observed that he was attended by a bird called ziczac. It is of the plover species, of a greyish colour, and as large as a small pigeon. The bird was walking up and down, close to the crocodile's nose. I suppose I moved, for suddenly it saw me, and instead of flying away, as any respectable bird would have done, he jumped up about a foot from the ground, screamed ziczac! with all the powers of his voice, and dashed himself against the crocodile's face two or three times. The great beast started up, and immediately spying his danger, made a jump into the air, and dashing into the water with a splash which covered me with mud, he dived into the river and disappeared. The ziczac, to my increased admiration, proud apparently of having saved his friend, remained walking up and down, uttering his cry, as I thought, with an exulting voice, and standing every now and then on his toes in a conceited manner, which made me justly angry with his impertinence. After having waited in vain for some time, to see whether the crocodile would come out again, I got up from the bank where I was lying, threw a clod of earth at the ziczac, and came back to the boat, feeling some consolation for the loss of my game, in having witnessed a circumstance, the truth of which has been disputed by several writers on natural history."

Pliny has described this crocodile-bird as a wren, but it is far more probable that it is a species of plover, the *Ammoptila charadrioides* of Swainson (*Pluvianus chlorocephalus*, Vieillot), and what it really does is doubtless to rid the crocodile of the swarms of flies and gnats which infest its palate to such a degree that the natural yellow colour is rendered black by them. In the same way we find the utmost harmony existing between sheep and cattle, and starlings, which perch upon their backs, and relieve them of the larvæ of insects deposited in their skins. So the rhinoceros bird is on terms of intimacy with the rhinoceros and hippopotamus, relieving them of insect pests, and by its watchful vigilance proving a most valuable sentinel. Mr. Gordon Cumming has described how his sport was spoiled by this bird, in the same way as Mr. Curzon's was spoiled by the impertinent ziczac.

Hybernation, or torpidity, so common in reptiles during the cold season, is participated in by the alligator tribe. It is said that the alligator of North America buries himself in the mud at the bottom of marshes till spring sets in, and that in severe frosts, animation is so completely suspended that slices can be cut from the animal without arousing it. On the other hand the alligator revels in the moist heat of Florida, and is especially formidable in numbers and dimensions at a mineral spring near the Mosquito river, where the water on issuing from the earth, is not only nearly boiling, but is strongly impregnated with copper and vitriol.

All writers agree as to the large number of alligators that infest the Amazon. The latest authority, Mr. Spruce, who is now engaged on an important botanical excursion in South America, writes thus of the Parana Miri dos Ramos.* "I was disappointed not to observe a single plant, save the rank grasses round the margin; but jacarés were laid in the water in almost countless numbers, resembling so many huge black stones or logs. What we had seen in the Amazon of these reptiles, was nothing compared to their abundance in the Ramos and its principal lakes. I can safely say, that at no one instant during the whole thirty days, when there was light enough to distinguish them, were we without one or more jacarés in sight."†

There are two species of these animals found here, one having a sharp mouth, the other a round one. The former grow to the length of about seven feet, and are called *jacaré-tingas*, or king jacarés. The other species attain the length of twenty-seven feet. In the inner lakes, towards the close of the rainy season, myriads of ducks breed in the rushes, and here the alligators swarm to the banquets of young birds. Mr. Edwards tells us, that should an adventurous sportsman succeed in arriving at one of these places, he has but a poor chance of bagging many

* An outlet from the Amazon.
† Hooker's Journal of Botany, Sept. 1851.

from the flocks around him; for the alligators are on the alert, and the instant a wounded bird strikes the water, they rush *en masse* for the poor victim, clambering over one another, and crashing their huge jaws upon each other's heads in the hasty seizure. The inhabitants universally believe that the alligator is paralyzed with fear at the sight of a tiger, and will suffer that animal to eat off its tail without making resistance. The story is complimentary to the tiger at all events, for the tail of the alligator is the only part in esteem by epicures.

The following incidents, which came under the immediate observation of Mr. Spruce, prove the ferocity of these fearful reptiles, which are the very scourge of the waters they infest.

Whilst in eager search after the *Victoria regia*, whose wonders have attracted so much interest at Kew and Chatsworth, Mr. Spruce was "glad to learn that it grew in a small lake on the opposite side of the Ramos, but I had no montaria to enable me to reach it, for one of our men, a Juma Indian, had run away a few nights previously, with our montaria and all our fishing-tackle, nor was there any montaria in the sitio where we were staying, but I was told I might borrow one at a sitio a little higher up. To this sitio I accordingly proceeded, and found at it an old man and his three sons, men of middle age, with their children. Two of the sons had just come in from a fishing expedition, the third had his arm in a sling,—and on inquiring the cause, I learnt that, seven weeks ago, he and his father had been fishing in the very lake I wished to visit, in a small montaria, and that having reached the middle and laid aside their paddles, they were waiting for the fish with their bows and arrows, when, unseen by them, a large jacaré glided under the montaria, gave them a jerk which threw them both into the water, and seizing the son by the right shoulder, dived with him at once to the bottom, the depth being, as they supposed, about four fathoms. In this fearful peril he had the presence of mind to thrust the fingers of his left hand into the monster's eyes, and after rolling over three or four times, the jacaré let go his hold, and the man rose to the surface, but mangled, bleeding, and helpless. His father immediately swam to his assistance, and providentially the two reached the shore without being further attacked. I was shown the wounds—*every tooth had told;* and some idea may be formed of this terrible gripe, when I state that the wounds inflicted by it extended from the collar-bone downwards to the elbow and the hip. All were now healed except one very bad one in the arm-pit, where at least one sinew was completely severed."

This was, Mr. Spruce remarks, no encouragement

to prosecute his enterprise, but being anxious to obtain the fruit of the Victoria, he was not the man to be deterred even by the prospect of a collision with these terrible jacarés; accordingly, as three of the little lads offered to row him over, he did not hesitate to avail himself of their services on the 21st of October, 1851. The outlet of the lake was speedily reached, when they disembarked and followed the dried bed of the igarapé, in which the lads were not slow to detect the recent footsteps of a jacaré. In five minutes more they reached the lake and embarked in the frail montaria, in which it was necessary so to place themselves before starting as to preserve an exact balance, and then they coasted along towards the Victoria, which appeared at the distance of some one hundred and fifty yards. " We had made but a few strokes when we perceived, by the muddy water ahead of us, that some animal had just dived. As we passed cautiously over the troubled water, a large jacaré came to the surface, a few yards from the off-side of our montaria, and then swam along parallel to our course, apparently watching our motions very closely. Athough the little fellows were frightened at the proximity of the jacaré, their piscatorial instincts were so strong, that at sight of a passing shoal of fish, they threw down their paddles and seized their mimic bows and arrows (the latter being merely strips of the leaf-stalk of the bacúba, with a few notches cut near the point), and one of them actually succeeded in piercing and securing an Aruana, of about eighteen inches long. Our scaly friend still stuck to us, and took no notice of our shouting and splashing in the water. At length the eldest lad bethought him of a large harpoon which was lying at the bottom of the montaria, He held this up and poised it in his hand, and the jacaré seemed at once to comprehend its use, for he retreated to the middle, and there remained stationary until we left the lake." Mr. Spruce was rewarded by finding three plants of the Victoria, of which one covered a surface of full six hundred square feet.

A singular fact, mentioned in Mr. Gosse's charming work, 'A Naturalist's Sojourn in Jamaica,' illustrates the predaceous vehemence and lurking patience of the alligator. In Spanish Haiti the large savanna rivers flow through wide, gently descending borders, carpeted with grass, having all the clean verdure of a lawn, and interspersed with clumps of beautiful flowering shrubs and trees. A Spanish priest, with three friends, had gone for a day's sporting to these grounds and had divided themselves. The three assembled at sunset, but the priest did not make his appearance. They sought him through the darkening thickets, and at last found him seated in a tree, into which he had been obliged to betake himself to escape an alligator that had pursued him by a succession of leaps. It had run in pursuit of him, as he said, jumping rapidly after him, with its back crooked like a frightened cat. He had taken refuge in the tree, whilst the reptile crouched in a thicket close by, quietly watching and waiting for his descent.

From an article on the Island of Bankar, in the 'Journal of the Indian Archipelago,' we learn that the alligator "is the most dangerous preying animal of the rivers. The river of Pankal Pinang is noted for the many victims of this monster. I (says the writer) often sat in the evening on the thick-wooded banks of the river of Batu Russak, and observed the motions of the crocodile (*Crocodilus biporcatus*). When the smooth river reflected in bright lustre the magnificent purple of heaven, when shadows overcast the waters with deeper darkness, and nothing interrupted the profound silence except the low bass of the Ludong, solemnly slow the alligator came on, and lifted his head above the water, forming nearly a right angle, with enormous jaws armed with pointed teeth, if opened. So it lay silent and terrific before me, its yellow eyes fixed upon me.

"In a work at present published by the Dutch naturalists, it is denied that the alligator has a voice. Humboldt admits it to have one, a roaring noise; but he himself never heard it. I once heard one make a

roaring noise while swimming in the river, opening its
jaws widely in doing so; and the fact has also been
testified to me by more than one hundred persons. I
threw cocoa-nuts into the stream to see whether the
crocodile was near: it came at once. I sent for the
dog of a soldier and threw it into the water, upon
which two large alligators, making a terrible roaring,
appeared on the spot with open jaws, and with long
leaps making at their prey. Passing the stream at
Idranayu, I found the remains of a Malay, who had
been devoured by an alligator. My colleague amputated, at Tobo-alli, the leg of a Chinese bit off by one.
My gardener lost one of his arms by an alligator
whilst fishing. In countries abounding in food it does
not attack men; as in several rivers of Celebes, where
the natives bathe at the places inhabited by the brutes,
and often mount on the backs of the crocodiles lying
at the bottom. If it is fed, it appears regularly at
the usual spot, and will come if called. The natives
like to dream of the crocodile, which they consider a
lucky omen: one however must not tell such a dream
to his mistress, as she would take it as a sign of infi-
delity."

Mr. Gosse mentions the following sad instance of
the ferocity of the crocodile. A young girl about
thirteen years of age, was washing a towel in the
Black River in St. Elizabeth's, in company with an

elder companion, at nightfall. She despised a warning to beware of the alligators, and just as she was boasting that she heeded no such danger, a scream for help, and a cry, "Lord have mercy upon me! Alligator has caught me!" apprised her companion, intent on her own washing, that the girl was carried off. The body was found some days after half-devoured; and two crocodiles, one nine feet long and the other seventeen, were hunted down and killed, with fragments of her body contained in them. There is a strange statement among the Negroes, that the manati, a cetacean inhabiting the same Black River, will remain watching a dead body if brought within its haunts, and, singular enough, the remainder of the body of this poor girl was found under the guardianship of a manati.

A writer in Silliman's 'American Journal' gives the following thrilling description of the capture and death of a huge alligator in one of the Philippine Islands:—" In the course of the year 1831, the proprietor of Halahala at Manilla, in the island of Luconia, informed me that he frequently lost horses and cows on a remote part of his plantation, and that the natives assured him that they were taken by an enormous alligator who frequented one of the streams which ran into the lake. Their descriptions were so highly wrought, that they were attributed to the fondness for exaggeration to which the inhabitants of that country are peculiarly addicted, and very little credit was given to their repeated relations. All doubts as to the existence of the animal were at last dispelled by the destruction of an Indian, who attempted to ford the river on horseback, although entreated to desist by his companions, who crossed at a shallow place higher up. He reached the centre of the stream and was laughing at the others for their prudence, when the alligator came upon him. His teeth encountered the saddle, which he tore from the horse, while the rider tumbled on the other side into the water and made for the shore. The horse, too terrified to move, stood trembling when the attack was made. The alligator, disregarding him, pursued the man, who safely reached the bank which he could easily have ascended, but, rendered foolhardy by his escape, he placed himself behind a tree which had fallen partly into the water, and drawing his heavy knife, leaned over the tree, and on the approach of his enemy struck him on the nose. The animal repeated his assaults and the Indian his blows, until the former, exasperated at the resistance, rushed on the man, and seizing him by the middle of the body, which was at once enclosed and crushed in his capacious jaws, swam into the lake. His friends hastened to the rescue, but the alligator slowly left the shore, while the poor wretch, writhing and shrieking in his agony, with his knife uplifted in his clasped hands, seemed, as the others expressed it, held out as a man would carry a torch. His sufferings were not long continued, for the monster sank to the bottom, and soon after reappearing alone on the surface, and calmly basking in the sun, gave to the horror-stricken spectators the fullest confirmation of the death and burial of their comrade.

"A short time after this event I made a visit to Halahala, and expressing a strong desire to capture or destroy the alligator, my host readily offered his assistance. The animal had been seen a few days before, with his head and one of his fore feet resting on the bank, and his eyes following the motions of some cows which were grazing near. Our informer likened his appearance to that of a cat watching a mouse, and in the attitude to spring upon his prey when it should come within his reach. I may here mention as a curious fact, that the domestic buffalo, which is almost continually in the water, and in the heat of day remains for hours with only his nose above the surface, is never molested by the alligator. All other animals become his victims when they incautiously approach him, and their knowledge of the danger most usually prompts them to resort to shallow places to quench their thirst.

"Having heard that the alligator had killed a horse,

we proceeded to the place, about five miles from the house. It was a tranquil spot, and one of singular beauty, even in that land. The stream, which a few hundred feet from the lake narrowed to a brook, with its green bank fringed with the graceful bamboo, and the alternate glory of glade and forest spreading far and wide, seemed fitted for other purposes than the familiar haunt of the huge creature that had appropriated it to himself. A few cane huts were situated at a short distance from the river, and we procured from them what men they contained, who were ready to assist in freeing themselves from their dangerous neighbour. The terror which he had inspired, especially since the death of their companion, had hitherto prevented them from making an effort to get rid of him, but they gladly availed themselves of our preparations, and, with the usual dependence of their character, were willing to do whatever example should dictate to them. Having reason to believe that the alligator was in the river, we commenced operations by sinking nets upright across its mouth, three deep, at intervals of several feet. The nets, which were of great strength, and intended for the capture of the buffalo, were fastened to trees on the banks, making a complete fence to the communication with the lake.

"My companion and myself placed ourselves with our guns on either side of the stream, while the Indians with long bamboos felt for the animal. For some time he refused to be disturbed, and we began to fear that he was not within our limits, when a spiral motion of the water under the spot where I was standing, led me to direct the natives to it, and the creature slowly moved on the bottom towards the nets, which he no sooner touched than he quietly turned back and proceeded up the stream. This movement was several times repeated, till, having no rest in the enclosure, he attempted to climb up the the bank. On receiving a ball in the body, he uttered a growl like that of an angry dog, and, plunging into the water, crossed to the other side, where he was received with a similar salutation, discharged directly into his mouth. Finding himself attacked on every side, he renewed his attempts to ascend the banks; but whatever part of him appeared was bored with bullets, and finding that he was hunted, he forgot his own formidable means of attack, and sought only safety from the troubles that surrounded him. A low spot which separated the river from the lake, a little above the nets, was unguarded, and we feared that he would succeed in escaping over it. It was here necessary to stand firmly against him, and in several attempts which he made to cross it, we turned him back with spears, bamboos, or whatever came first to hand. He once seemed determined to force his way,

and foaming with rage, rushed with open jaws and
gnashing his teeth with a sound too ominous to be despised, appeared to have his full energies aroused, when
his career was stopped by a large bamboo thrust violently into his mouth, which he ground to pieces, and
the fingers of the holder were so paralyzed that for
some minutes he was incapable of resuming his gun.
The natives had now become so excited as to forget
all prudence, and the women and children of the little
hamlet had come down to the shore to share in the
general enthusiasm. They crowded to the opening,
and were so unmindful of their danger that it was
necessary to drive them back with some violence.
Had the monster known his own strength and dared
to have used it, he would have gone over that spot with
a force which no human power could have withstood,
and would have crushed or carried with him into the
lake about the whole population of the place. It is
not strange that personal safety was forgotten in the
excitement of the scene. The tremendous brute,
galled with wounds and repeated defeat, tore his way
through the foaming water, glancing from side to side
in the vain attempt to avoid his foes; then rapidly
ploughing up the stream he grounded on the shallows,
and turned back frantic and bewildered at his circum-
scribed position. At length, maddened with suffering
and desperate from continued persecution, he rushed

furiously to the mouth of the stream, burst through two of the nets, and I threw down my gun in despair, for it looked as though his way at last was clear to the wide lake; but the third net stopped him, and his teeth and legs had got entangled in all. This gave us a chance of closer warfare with lances, such as are used against the wild buffalo. We had sent for this weapon at the commencement of the attack, and found it much more effectual than guns. Entering the canoe, we plunged lance after lance into the alligator, as he was struggling under the water, till a wood seemed growing from him, which moved violently above while his body was concealed below. His endeavours to extricate himself lashed the waters into foam mingled with blood, and there seemed no end to his vitality or decrease to his resistance, till a lance struck him directly through the middle of the back, which an Indian, with a heavy piece of wood, hammered into him as he could catch an opportunity. My companion on the other side now tried to haul him to the shore, by the nets to which he had fastened himself, but had not sufficient assistance with him. As I had more force with me, we managed, by the aid of the women and children, to drag his head and part of his body on to the little beach, and giving him the *coup de grâce*, left him to gasp out the remnant of his life."

This monster was nearly thirty feet in length and

thirteen feet in circumference, and the head alone weighed three hundred pounds. On opening him there were found, with other parts of the horse, three legs entire, torn off at the haunch and shoulder, besides a large quantity of stones, some of them several pounds' weight.

The flesh of alligators is eaten by some nations, but can scarcely be considered an epicurean morsel. A serpent will, however, lunch off an alligator with infinite *gusto*, as appears from the following example. In October, 1822, a Camoudy snake was killed in Demerara, measuring fourteen feet long and thirteen inches in circumference, as the natural size of the body, but the belly was distended to the enormous size of thirty-one inches. On opening it, it was found to contain an entire alligator, recently swallowed, and measuring six feet long by twenty-eight inches in circumference. From the appearance of the neck of the alligator, it was evident that the snake destroyed him by twining round that part; and so severe had been the constriction that the eyes were starting from the head.

Some valuable information as to the habits of crocodiles is given by Mr. Gosse, on the authority of Mr. Richard Hill. It is generally supposed that alligators are greedily partial to dogs, and surprise them often when they come to drink at the river. "The voice of the dog," says Mr. Hill, "will always draw them away from an object when prowling. Those who would cross a river without any risk from their attacks, send a scout down the stream to imitate the canine bark, yelp, or howl, when away swim the alligators for their prey, leaving an unmolested ford for the traveller higher up. Instinct has taught the dog to secure himself by a similar expedient. When it has to traverse a stretch of water, it boldly goes some distance down the stream, and howls and barks. On perceiving the alligators crowding in eager cupidity to the spot, it creeps gently along the banks higher up, and swims over the water without much fear of being pursued." Mr. Hill, however, doubts whether this eagerness with which the alligator responds to the cry of the dog is to be attributed to fondness for it as food; he rather ascribes it to the similarity of the sound to its own peculiar cry under any species of excitement, especially the impassioned voice of its young, and he considers that the creatures press towards the point whence the cry is heard, the females to protect the young, the males to devour them.

Sir Hans Sloane, in his 'Natural History of Jamaica,' gives the following curious account of the mode of taking alligators there practised:—

"They are very common on the coasts and deep rivers of Jamaica. One of nineteen feet in length,

I was told, was taken by a dog, which was made use of as a bait, with a piece of wood tied to a cord, the farther end of which was fastened to a bed-post. The crocodile coming round as usual every night, seized the dog, was taken by the piece of wood made fast to the cord, drew the bed to the window and waked the people, who killed the alligator, which had done them much mischief. The skin was stuffed and offered to me as a rarity and present, but I could not accept of it, because of its largeness, wanting room to stow it."

Few authors have received greater injustice than Mr. Waterton, on account of an adventure described by him in his interesting 'Wanderings.' "Having caught (he says), by means of a baited hook, a huge cayman in the river Essequibo, they (the Indians) pulled the cayman within two yards of me; I saw he was in state of fear and perturbation. I instantly dropped the mast, sprang up and jumped on his back, turning half round as I vaulted, so that I gained my seat with my face in a right position. I immediately seized his fore-legs, and, by main force, twisted them on his back: thus they served me for a bridle."

This statement excited a whirlwind of ridicule; we will see with how little cause or justice.

When speaking of the Tentyrites, we have quoted ancient authors proving that it was the usual proceeding of these people to spring on the backs of crocodiles, and in that position to subdue them. Dr. Pocock, in his 'Observations on Egypt,' describes the following method of taking the crocodile:—"The inhabitants," says he, "make some animal cry at a distance from the river, and when the crocodile comes out, they thrust a spear into his body, to which a rope is tied: they then let him go into the water to spend himself, and afterwards drawing him out, run a pole into his mouth, and jumping on his back, tie his jaws together." Such is the mode still practised in Egypt, and the following interesting account given by Mr. Gosse, of the capture of an alligator in Jamaica, most fully exonerates Mr. Waterton from any suspicion of exaggeration:—

"A cayman from the neighbouring lagoons of Lyson's estate in St. Thomas's in the East, that used occasionally to poach the ducks and ducklings having free warren about the water-mill, was taken in his prowl and killed. All sorts of suspicion was entertained about the depredators among the ducks, till the crocodile was surprised lounging in one of the ponds, after a night's plunder. Downie, the engineer of the plantation, shot at him and wounded him, and though it did not seem that he was much hurt, he was hit with such sensitive effect, that he immediately rose out of the pond to regain the morass. It was now that David Brown, an African wainman, came up, and before the reptile could make a dodge to get away, he threw himself astride over his back, snatched up his fore paws in a moment, and held them doubled up. The beast was immediately thrown upon his snout, and though able to move freely his hind feet, and slap his tail about, he could not budge half a yard, his power being altogether spent in a fruitless endeavour to grub himself onward. As he was necessarily confined to move in a circle, he was pretty nearly held to one spot. The African kept his seat. His place across the beast being at the shoulders, he was exposed only to severe jerks, as a chance of being thrown off. In this way a huge reptile eighteen feet long,—for so he measured when killed, —was held *manu forti*, by one man, till Downie reloaded his fowling-piece, and shot him quietly through the brain."

We will, finally, see how our bold British subalterns in India, when in steeple-chasing mood, deal with an alligator steed. Their proceedings are thus described by Lieutenant Burton.

"The 'Alligator Tank,' as it is called by the natives, owes its origin and fame to one Hajee Mufur, a Moslem hermit, who first visited the barren spot, and to save himself the trouble of having to fetch water from afar, caused a rill to trickle from the rock above. It was visited by four brother saints, who, without rhyme or reason, began to perpetrate a variety of miracles. One formed a hot mineral spring, whose graveolent proceeds settled in the nearest hollow, converting it into a foul morass; another metamorphosed a flower into an animal of the crocodile species; and a third converted the bit of stick he was wont to use as a tooth-brush, into a palm-shoot, which at once becoming a date-tree, afforded the friends sweet fruit and pleasant shade! This spot, and the inhabitants of the morass, descendants of the floral reptilian, are regarded as holy by the natives, but are subjected to much persecution from the youthful officers of the British army, as they are politely called. One of these having performed the feat of running across the morass unharmed, and being in a state of great pale-alean valour, proposes an alligator-ride, is again laughed to scorn, and again runs off, with mind made up, to the tent. A moment afterwards he reappears, carrying a huge steel fork, and a sharp hook, strong and sharp, with the body of a fowl quivering on one end, and a stout cord attached to the other; he lashes his line carefully round one of the palm-trees, and commences plying the water for an alligator.

"A brute, nearly twenty feet long, a real Saurian every inch of him, takes the bait and finds himself in a predicament; he must either disgorge a savoury morsel, or remain a prisoner, and for a moment or two he makes the ignoble choice. He pulls, however, like a thorough-bred bulldog, shakes his head as if he wished to shed it, and lashes his tail with the energy of a shark who is being beaten to death with capstan-bars. In a moment young Waterton is seated, like an elephant-driver, upon the thick neck of the reptile, who, not being accustomed to carry such weight, at once sacrifices his fowl, and, running off with his rider, makes for the morass. On the way, at times, he slackens his zigzag wriggling course, and attempts a bite, but the prongs of the steel fork, well rammed into the soft skin of his neck, muzzle him effectually enough; and just as the steed is plunging into his own element, the jockey springs actively up, leaps on one side, avoids a terrific lash from the serrated tail, and again escapes better than he deserves."*

With the exception of the bird already mentioned, the crocodilians may be said to be at war with every creature that crosses their path; yet they play their part, and that an important one, in the economy of Nature. They are to the great rivers of the tropics what the wolves and hyænas are to the land, and the sharks to the sea—scavengers clearing away offal and carrion, which would poison the waters not less than taint the air. Fortunate is it that such a diversity in

* From 'Scinde, or the Unhappy Valley.' Bentley, 1851.

the choice of food exists! That which is disgusting to one class of animals is sought after with avidity by another; and thus, apart from every other consideration, Providence has, by establishing a variety of tastes, prevented the possibility of such a dearth of sustenance as might have arisen if all tribes had resembled one another in the choice and character of their food.

www.ingramcontent.com/pod-product-compliance
Lightning Source LLC
Chambersburg PA
CBHW030313240426
43673CB00040B/1147